Holt McDougal Mathematics

Course 2
Chapter 3 Resource Book

HOLT McDOUGAL
a division of Houghton Mifflin Harcourt

Printed in the United States of America

ISBN 13: 978-0-55-400728-1
ISBN 10: 0-55-400728-2

1 2 3 4 5 170 12 11 10 09 08

Contents

Section 3A Family Letter 1
Section 3A At-Home Practice 3
Section 3A Family Fun 4
3-1 Practice A, B, C 5
3-1 Review for Mastery............................ 8
3-1 Challenge 9
3-1 Problem Solving 10
3-1 Reading Strategies...........................11
3-1 Puzzles, Twisters & Teasers 12
3-2 Practice A, B, C 13
3-2 Review for Mastery.......................... 16
3-2 Challenge 17
3-2 Problem Solving 18
3-2 Reading Strategies.......................... 19
3-2 Puzzles, Twisters & Teasers 20
3-3 Practice A, B, C 21
3-3 Review for Mastery.......................... 24
3-3 Challenge 25
3-3 Problem Solving 26
3-3 Reading Strategies.......................... 27
3-3 Puzzles, Twisters & Teasers 28
3-4 Practice A, B, C 29
3-4 Review for Mastery.......................... 32
3-4 Challenge 34
3-4 Problem Solving 35
3-4 Reading Strategies.......................... 36
3-4 Puzzles, Twisters & Teasers 37
3-5 Practice A, B, C 38
3-5 Review for Mastery.......................... 41
3-5 Challenge 43
3-5 Problem Solving 44
3-5 Reading Strategies.......................... 45
3-5 Puzzles, Twisters & Teasers 46
Section 3B Family Letter 47
Section 3B At-Home Practice 49
Section 3B Family Fun 50
3-6 Practice A, B, C 51
3-6 Review for Mastery.......................... 54
3-6 Challenge 55
3-6 Problem Solving 56
3-6 Reading Strategies.......................... 57
3-6 Puzzles, Twisters & Teasers 58
3-7 Practice A, B, C 59
3-7 Review for Mastery.......................... 62
3-7 Challenge 63
3-7 Problem Solving 64
3-7 Reading Strategies.......................... 65
3-7 Puzzles, Twisters & Teasers 66
3-8 Practice A, B, C 67
3-8 Review for Mastery.......................... 70
3-8 Challenge 72
3-8 Problem Solving 73
3-8 Reading Strategies.......................... 74
3-8 Puzzles, Twisters & Teasers 75
3-9 Practice A, B, C 76
3-9 Review for Mastery.......................... 79
3-9 Challenge 80
3-9 Problem Solving 81
3-9 Reading Strategies.......................... 82
3-9 Puzzles, Twisters & Teasers 83
3-10 Practice A, B, C 84
3-10 Review for Mastery........................ 87
3-10 Challenge 88
3-10 Problem Solving 89
3-10 Reading Strategies........................ 90
3-10 Puzzles, Twisters & Teasers 91
3-11 Practice A, B, C 92
3-11 Review for Mastery........................ 95
3-11 Challenge 97
3-11 Problem Solving 98
3-11 Reading Strategies........................ 99
3-11 Puzzles, Twisters & Teasers 100
Answers 101

Description of Contents

Family Involvement Pages

The Chapter Resource Book includes a set of family involvement pages for each section. The family involvement pages consist of the following items:

- **Family Letter**, a two-page letter describing the math that the student will study in each section. A list of vocabulary words from the lessons is included, and some worked-out examples are provided.
- **At-Home Practice**, a one-page worksheet with problems drawn from the content described in the Family Letter. Answers are provided on the page so that the student and his or her family can check this work at home.
- **Family Fun**, a one-page activity sheet that the student and his or her family can work on together.

Practice A, B, and C

There are three practice worksheets for every lesson. All of these reinforce the content of the lesson. Practice B is shown in the Teacher's Edition and is appropriate for the on-level student. It is also available as a workbook (Homework & Practice Workbook)

Practice A is easier than Practice B but still practices the content of the lesson. Practice C is more challenging than Practice B.

Review for Mastery

The Review for Mastery worksheet (one per lesson) provides an alternate way to teach or review the main concepts of the lesson. This worksheet is one or two pages long and is shown in the Teacher's Edition.

Challenge

The Challenge worksheet (one per lesson) enhances critical thinking skills and extends the lesson. This worksheet is shown in the Teacher's Edition.

Problem Solving

The Problem Solving worksheet (one per lesson) provides practice in problem solving and opportunities for real-world applications and for interdisciplinary connections. There are both multiple choice and short response problems. This worksheet is shown in the Teacher's Edition.

Reading Strategies

The Reading Strategies worksheet (one per lesson) provides tools to help the student master math vocabulary or symbols.

Puzzles, Twisters & Teasers

The Puzzles, Twisters & Teasers worksheet (one per lesson) provides fun practice while reinforcing the content of the lesson.

CHAPTER 3

Family Letter

3A Decimal Operations and Applications

Dear Family,

The student will be learning how to apply the four basic mathematical operations, addition, subtraction, multiplication, and division, to decimals. This process will begin by first learning how to estimate the result of a decimal problem. This is an important step because it teaches the student how to recognize whether the solution is reasonable.

The student will estimate sums and differences of decimals by rounding to the nearest integer.

Estimate by rounding to the nearest integer.
67.8 + 43.1

67.8	→	68	8 > 5, so round to 68.
+ 43.1	→	+ 43	1 < 5, so round to 43.
		111	← Estimate.

When estimating products or quotients, the student will use compatible numbers to find the most reasonable estimate. A compatible number is a number that can replace the decimal in the problem in order to make the problem easier to solve.

When multiplying, round the numbers to the nearest nonzero number or to numbers that are easy to multiply.

Estimate using compatible numbers.
23.88 × 4.89

23.88	→	25	25 is easy to multiply.
× 4.89		× 5	8 > 5, so round to 5.
		125	← Estimate.

When estimating a quotient, round the numbers so that the quotient will not have a remainder.

Estimate using compatible numbers.
47.8 ÷ 6.02

47.8 ÷ 6.02

47.8 → Round to 48.

6.02 → 6 divides evenly into 48.

48 ÷ 6 = 8 ← Estimate.

Vocabulary

These are the math words we are learning:

compatible numbers numbers that can replace the numbers in the problem and are easy to use

Family Letter

3A Decimal Operations and Applications continued

When adding and subtracting decimals, the student will learn to line up the decimal points and then add or subtract as usual.

Add. Estimate to check whether your answer is reasonable.
57.68 + 16.02

57.68	Line up the decimal points.
+ 16.02	
73.70	Add.

Estimate
58 + 16 = 74   73.70 is a reasonable answer.

The student will find products and quotients of decimals the same way as multiplying and dividing integers. He or she will learn where to properly place the decimal point and how to determine if the answer is reasonable.

Multiply Integers by Decimals	**Multiply Decimals by Decimals**
8 × 0.4 Add the number of decimal places in each number multiplied. 8 → 0 decimal place 0.4 → + 1 decimal place 1 decimal place Multiply. Place the decimal point **1** digit to the left. 8 × 0.4 = 3.2	**0.8 × 0.4** Add the number of decimal places in each number multiplied. 0.8 → 1 decimal place 0.4 → + 1 decimal place **2** decimal place Multiply. Place the decimal point **2** digit to the left. 0.8 × 0.4 = 0.32
Divide Decimals by Integers **18.81 ÷ 3** $$\begin{array}{r} 6.27 \\ 3\overline{)18.81} \\ -18 \\ \hline 8 \\ -6 \\ \hline 21 \\ -21 \\ \hline 0 \end{array}$$ Place the decimal point in the quotient directly above the decimal point in the dividend.	**Divide Decimals by Decimals** **18.81 ÷ 0.03** $$\begin{array}{r} 627 \\ 3\overline{)1881} \\ -18 \\ \hline 8 \\ -6 \\ \hline 21 \\ -21 \\ \hline 0 \end{array}$$ Multiply both numbers by 100. Divide as you would with whole numbers.

Sincerely,

Name ______________________ Date ______________ Class ______________

CHAPTER 3

At-Home Practice

3A Decimal Operations and Applications

Estimate by rounding.

1. 3.29 + 71.968 ____________
2. 45.125 – 27.5 ____________
3. –5.48 + 32.664 ____________

Estimate by using compatible numbers.

4. 58.82 ÷ 6.41 ____________
5. –7.94 • 62.145 ____________
6. 24.18 • (–5.53) ____________

Add. Estimate to check whether your answer is reasonable.

7. 1.36 + 74.76 ____________
8. –6.63 + (–32.27) ____________
9. 19 + 8.98 ____________

Subtract. Estimate to check whether your answer is reasonable.

10. 36 – 7.63 ____________
11. 9.42 – 0.86 ____________
12. 25.3 – 4.53 ____________

Multiply. Estimate to check whether your answer is reasonable.

13. 8.5 • 3.4 ____________
14. –5.61 • 0.7 ____________
15. 6.92 • 2.4 ____________

Divide. Estimate to check whether your answer is reasonable.

16. (–58.31) ÷ 4.9 ____________
17 0.217 ÷ 7 ____________
18. 42 ÷ (–3.5) ____________

Solve.

19. $3.68 + a = -21.37$ ____________
20. $x - 8.21 = 34.87$ ____________
21. $\frac{w}{5.7} = 6.3$ ____________

Answers: **1.** 75 **2.** 17 **3.** 28 **4.** 10 **5.** –480 **6.** –125 **7.** 76.12 **8.** –38.9 **9.** 27.98 **10.** 28.37 **11.** 8.56 **12.** 20.77 **13.** 28.9 **14.** –3.927 **15.** 16.608 **16.** –11.9 **17.** 0.031 **18.** –12 **19.** $a = -25.05$ **20.** $x = 43.08$ **21.** w = 35.91

Name ______________________ Date ______________ Class ______________

CHAPTER 3

Family Fun

Quotable Quotes

Directions

The product is given in the first column. Find the correct problem for each product in the shaded columns. Then write the letter of the answer on the lines below to solve the puzzle.

1. 0.0672	**R** 2.8 × 0.24	**S** 0.28 × 2.4	**T** 0.028 × 2.4
2. 0.00385	**H** 0.005 × 0.77	**I** 0.05 × 7.7	**J** 0.05 × 0.77
3. 2.6656	**M** 39.2 × 6.8	**N** 3.92 × 6.8	**O** 3.92 × 0.68
4. 0.00027	**L** 0.03 × 0.09	**M** 0.003 × 0.09	**N** 0.003 × 0.9
5. 2.09709	**A** 8.63 × 0.243	**B** 86.3 × 24.3	**C** 8.63 × 2.43
6. 277.178	**R** 29.3 × 94.6	**S** 2.93 × 94.6	**T** 293 × 9.46
7. 1.2789	**E** 18.27 × 0.07	**F** 1.827 × 0.07	**G** 0.1827 × 0.07
8. 238.018	**B** 34.9 × 68.2	**C** 3.49 × 6.82	**D** 3.49 × 68.2
9. 0.49928	**I** 0.079 × 6.32	**J** 0.79 × 6.32	**K** 0.79 × 63.2
10. 0.5133	**S** 0.59 × 0.87	**T** 0.059 × 0.087	**U** 5.9 × 8.7
11. 19.64928	**M** 2.04 × 96.32	**O** 0.204 × 96.32	**N** 20.4 × 96.32
12. 44.436	**N** 4.83 × 9.2	**O** 48.3 × 9.2	**P** 4.83 × 0.92

Which genius made this famous remark?

Genius is one percent inspiration and ninety-nine percent perspiration.

___ ___ ___ ___ ___ ___   ___ ___ ___ ___ ___ ___
1. 2. 3. 4. 5. 6.   7. 8. 9. 10. 11. 12.

Answer: Thomas Edison

Name ________________ Date ________ Class ________

LESSON 3-1

Practice A

Estimating with Decimals

Round to the nearest whole number.

1. 4.23 ________
2. 1.91 ________
3. 10.75 ________
4. 5.88 ________
5. 12.07 ________
6. 18.70 ________

Estimate by rounding.

7. 8.4 + 15.9 ________
8. 3.45 + 5.34 ________
9. 9.36 + 7.542 ________
10. 9.6 + (−7.2) ________
11. −99.67 + −49.87 ________
12. −4.5 + 6.2 ________
13. 16.8 − 4.3 ________
14. 34.25 − 18.5 ________
15. 23.3 − 3.66 ________

Estimate using compatible numbers.

16. 7.9 • 0.6 ________
17. 6.35 • 1.83 ________
18. 4.4 • 7.15 ________
19. 5.24 • 10.78 ________
20. 29.1 • 5.15 ________
21. 14.25 • 2.99 ________
22. 16.2 ÷ 3.5 ________
23. 32.5 ÷ 3.2 ________
24. 36.34 ÷ 2.1 ________
25. Cheri stopped for gasoline at a station that was charging $2.39 per gallon. If Cheri had $12.45 in cash, about how many gallons of gas could she buy? ________

Name ________________ Date ________ Class ________

LESSON 3-1

Practice B

Estimating with Decimals

Estimate by rounding to the nearest integer.

1. 7.45 + 35.84 ________
2. 64.08 – 23.47 ________
3. 6.842 + 14.05 ________
4. 7.156 + 8.34 ________
5. 84.23 + (–78.24) ________
6. 3.78 – 2.078 ________
7. 46.47 – 98.75 ________
8. 87.24 – 56.38 ________
9. 6.324 + 60.324 ________
10. –28.318 + 18.955 ________
11. 35.082 + 8.37 ________
12. –62.49 – 12.84 ________

Use compatible numbers to estimate.

13. 59.69 ÷ 19.904 ________
14. 86.234 • 9.876 ________
15. 54.87 • 19.47 ________
16. –16.04 • 10.45 ________
17. 31.25 • 6.57 ________
18. 92.67 ÷ 32.89 ________
19. 5.548 • 12.38 ________
20. 88.42 ÷ 7.589 ________
21. 90.05 ÷ 6.21 ________

22. Lisha works 20 hours per week at the bowling alley and makes $8.55 an hour. She gets a raise of $1.30 an hour. Approximately how much more will she make each week with her raise?

____

23. Miguel is able to save $87.34 each month. He wants to buy a guitar that costs $542.45. For about how many months will Miguel have to save before he can buy the guitar?

____

Name ______________________ Date ______________ Class ______________

LESSON 3-1

Practice C

Estimating with Decimals

Estimate.

1. 6.187 + 61.871 ________

2. 52.846 − (−5.284) ________

3. 77.435 ÷ 21.234 ________

4. 7.894 + (−7.894) ________

5. 84.324 • 9.846 ________

6. 47.445 + 96.845 ________

7. −896.23 − 24.571 ________

8. −7.501 • 67.499 ________

9. 483.88 ÷ (−15.73) ________

10. 804.504 + 905.405 ________

11. −46.781 − 5.23 ________

12. 30.214 • 89.321 ________

13. 98.346 + 162.54 ________

14. 84.067 + (−101.26) ________

15. 720.089 ÷ 44.506 ________

16. 645.645 • (−1.49) ________

17. 563.647 ÷ 27.804 ________

18. −61.248 + (−65.234) ________

19. 2.605 − 11.019 ________

20. 40.93 • 15.042 ________

21. 27.38 − (−6.27) ________

22. 125.88 ÷ 23.91 ________

23. −38.5 ÷ 5.34 ________

24. 59.64 • (−11.6) ________

25. Niles is driving from New York City to Baltimore. He drives 64.8 mi/h for 2.9 hours. Approximately how far is Baltimore from New York City?

____

26. Charlotte had $204. She bought a skirt for $78 and a scarf for $59. She wants to buy a pair of boots that costs $75. Does she have enough money left to buy the boots? About how much money does she have?

____

Name ______________________ Date ______________ Class ______________

LESSON 3-1

Review for Mastery

Estimating with Decimals

You can estimate with decimals by rounding each number to its greatest **place.**

Estimate: 52.38 + 9.006 The greatest place of 52.38 is **tens.** The greatest place of 9.006 is **ones.** 52.38 → 50 9.006 → 9 50 + 9 = 59 So, 52.38 + 9.006 is about 59.	**Estimate: 97.45 − 14.9** The greatest place of 97.45 is **tens.** The greatest place of 14.9 is **tens.** 97.45 → 100 14.9 → 10 100 − 10 = 90 So, 97.45 − 14.9 is about 90.
Estimate: 16.35 • 1.8 The greatest place of 16.35 is **tens.** The greatest place of 1.8 is **ones.** 16.35 → 20 1.8 → 2 20 • 2 = 40 So, 16.35 • 1.8 is about 40.	**Estimate: 29.7 ÷ 3.65** The greatest place of 29.7 is **tens.** The greatest place of 3.65 is **ones.** 29.7 → 30 3.65 → 4 30 ÷ 4 is about 32 ÷ 4 = 8 So, 29.7 ÷ 3.65 is about 8.

Estimate.

1. 7.843 + 54.1

   Greatest place of 7.843 __________

   7.843 rounds to __________

   Greatest place of 54.1 __________

   54.1 rounds to __________

   Estimate: ____________________

2. 28.45 − 14.602

   Greatest place of 28.45 __________

   28.45 rounds to __________

   Greatest place of 14.602 __________

   14.602 rounds to __________

   Estimate: ____________________

3. 6.41 • (−19.725)

   Greatest place of 6.41 __________

   6.41 rounds to __________

   Greatest place of −19.75 __________

   −19.725 rounds to __________

   Estimate: ____________________

4. 44.67 ÷ 8.6

   Greatest place of 44.67 __________

   44.67 rounds to __________

   Greatest place of 8.6 __________

   8.6 rounds to __________

   __________ is about __________

   Estimate: ____________________

5. 36.5 + 78.09 ______________

6. 9.45 + (−2.75) ______________

7. 98.56 − 53.381 ______________

8. 33.52 • 5.29 ______________

9. −68.3 • 4.344 ______________

10. 29.65 ÷ 5.92 ______________

Name ______________________ Date ______________ Class ______________

LESSON 3-1

Challenge

Just Estimate

Use estimates to compare: $8.62 - 15.91$ [?] $-2.4(3.8)$

$8.62 - 15.91$ [?] $-2.4(3.8)$

Think: $9 - 16 = -7$ **Think:** $-2 \cdot 4 = -8$

$-7 > -8$, so $8.62 - 15.91 > -2.4(3.8)$

Estimate to compare. Use > or <.

1. $3.45 + 11.65$ ___ $16.8 - 4.1$
2. $-0.54(14.22)$ ___ $-23.65 + 9.44$
3. $-2.59 - 5.18$ ___ $6.12 - 13.3$
4. $-9.6(-7.25)$ ___ $82.45 - 10.3$
5. $24.7 \div 4.5$ ___ $12.8 - 5.3$
6. $-20.8 + 12.3$ ___ $-20.2 \div 1.8$
7. $125.95 - 25.4$ ___ $1.5(50.38)$
8. $-14.57 - 7.34$ ___ $-39.62 \div 2.3$
9. $4.9(-11.6)$ ___ $40.29 + 20.51$
10. $47.6 - 17.7$ ___ $-7.7(-4.1)$
11. $23.6 \div 5.5$ ___ $13.08 - 8.4$
12. $9.83 - 23.41$ ___ $-6.42 - 5.19$
13. $18.45 \div 1.82$ ___ $49.5 \div 5.46$
14. $0.35(-3.55)$ ___ $4.68 - 5.54$
15. $32.7 \div (-2.6)$ ___ $-5.28(1.52)$
16. $-5.49(-4.06)$ ___ $24.62 - 6.4$
17. $-3.24 - 16.4$ ___ $-6.6(2.8)$
18. $0.78(56.5)$ ___ $-9.1 + 66.65$
19. $-45.3 \div 4.55$ ___ $79.9 \div (-8.48)$
20. $-6.7(-5.08)$ ___ $18.5(1.54)$
21. $65.36 \div (-10.7)$ ___ $-9.3 + 1.48$
22. $-3.53(-2.45)$ ___ $-119.5 \div (-11.7)$
23. $65.5 - 67.09$ ___ $34.4 \div 33.55$
24. $-40.06(-3.1)$ ___ $97.8 + 19.7$

Name ______________________ Date ______________ Class ______________

LESSON 3-1

Problem Solving

Estimating with Decimals

Write the correct answer.

1. The Spanish Club makes a profit of $1.85 on every pie sold at a bake sale. The goal is to earn $40.00 selling pies. Will the club have to sell more than or fewer than 20 pies to meet the goal?

____

2. Murray and 4 friends split the cost of a pizza, with each paying the same amount. Murray has $3.36. The pizza costs $14.20. Does Murray have enough to pay for his share? About how much is his share?

____

3. Luis has 22 MB of free space on his MP3 player. He wants to download 5 songs. They will take up 5.1, 4.1, 4.3, 8.2, and 3.6 MB of space. Does Luis have enough free space? About how much space is needed?

____

4. Debi has a gift certificate worth $50 for a local book-and-music store. She decides to buy a book that costs $18.95 and some CDs. Each CD costs $13.99. How many CDs can she buy?

____

Choose the letter for the best answer.

This table shows the number of dollars foreign tourists spent while visiting different countries in 2003.

Spending by Tourists in 2003

Country	Amount Spent by Foreign Tourists ($ billions)
United States	65.1
Spain	41.7
France	36.3
Italy	31.3
Germany	22.8
United Kingdom	19.5
China	17.4
Austria	13.6
Turkey	13.2
Greece	10.6
Mexico	9.5

5. The amount spent in which two countries was about equal to the amount spent in Italy?
   A Austria and Turkey
   B United Kingdom and China
   C Germany and Greece
   D Austria and China

6. About how much did tourists spend in Spain, France, and Italy all together?
   F $111 billion
   G $109 billion
   H $78 billion
   J $67 billion

7. If the amount spent by foreign tourists remains the same, about how much will foreign tourists spend in the United States over 5 years?
   F $120 billion
   G $250 billion
   H $350 billion
   J $450 billion

Name ______________________ Date ______________ Class ______________

LESSON 3-1

Reading Strategies

Make Generalizations

Estimating is useful when you don't need an exact answer. Knowing how to round decimals helps you estimate.

Rules for Rounding Decimals
1. Look at the digit in the tenths place.
2. If that digit is five or greater, round the number in the ones place up one.
3. If that digit is less than five, round the number in the ones place down one.

Estimate the sum. Use the rules in the table to answer each question.

$$\begin{array}{r} 7.48 \\ +\ 2.68 \\ \hline \end{array}$$

1. In the number 7.48, will you round 7 up or down? Explain.

____

2. In the number 2.68, will you round 2 up or down? Explain.

____

3. Rewrite the equation and solve using the rounded numbers.

____

Use the rounding rules to estimate the difference.

$$\begin{array}{r} 13.78 \\ -\ 5.23 \\ \hline \end{array}$$

4. What is 13.78 rounded to the nearest whole number?

____

5. What is 5.23 rounded to the nearest whole number?

____

6. Is it easier to solve 13.78 – 5.23 or 14 – 5? Explain.

____

Name ______________________ Date ______________ Class ______________

LESSON 3-1

Puzzles, Twisters & Teasers

Round and Round We Go!

Estimate sums and differences by rounding. Estimate products and quotients by using compatible numbers. Then solve the riddle.

I	−9.916 + 12.4	________
D	24.79 • 9.83	________
L	37.2 + 25.83	________
N	−36.8 + 14.217	________
G	68.2 + 23.67	________
U	61.45 ÷ 9.08	________
O	15 − 6.835	________
R	37.63 ÷ 7.43	________
A	5.921 − 13.2	________
H	98.6 + 43.921	________
E	62.84 − 35.169	________
Y	−3.96 • 14.81	________

What did the hat say to the tie?

___ ___ ___ , ___ ___ ___ ___
2 63 63 92 8 8 −23

___ ___ ___ ___ ___ , ___ ___ ___
−7 143 28 −7 250 −60 8 7

___ ___ ___ ___
143 −7 −23 92

___ ___ ___ ___ ___ ___ .
−7 5 8 7 −23 250

Name ______________________ Date ______________ Class ______________

LESSON 3-2

Practice A

Adding and Subtracting Decimals

Add. Estimate to check whether your answer is reasonable.

1. 3.52 + 6.33

	Tens	Ones	.	Tenths	Hundredths
		3	.	5	2
+		6	.	3	3
			.		

2. 9.48 + 12.75

	Tens	Ones	.	Tenths	Hundredths
		9	.	4	8
+	1	2	.	7	5
			.		

3. 7.97 + 3.6

	Tens	Ones	.	Tenths	Hundredths
		7	.	9	7
+		3	.	6	
			.		

4. 13 + (−10.35)

	Tens	Ones	.	Tenths	Hundredths
	1	3	.		
+	−1	0	.	3	5
			.		

5. 0.82 + 1.54 ______________

6. 12.28 + 19.019 ______________

7. 29.5 + 3.676 ______________

Subtract. Estimate to check whether your answer is reasonable.

8. 24.6 − 17.93

	Tens	Ones	.	Tenths	Hundredths
	2	4	.	6	
−	1	7	.	9	3
			.		

9. 40 − 19.4

	Tens	Ones	.	Tenths	Hundredths
	4	0	.		
−	1	9	.	4	
			.		

10. 8.4 − 4.6 ______________

11. 9.2 − 3.8 ______________

12. 13.4 − 8.2 ______________

13. 15.42 − 8.3 ______________

14. 9 − 5.5 ______________

15. 21.68 − 12 ______________

Name ______________________ Date ____________ Class ____________

LESSON 3-2

Practice B

Adding and Subtracting Decimals

Add. Estimate to check whether each answer is reasonable.

1. 6.14 + 8.91 ________
2. 4.51 + 13.08 ________
3. 12.54 + 21.08 ________
4. 34.22 + (–18.5) ________
5. –10.10 + (–5.9) ________
6. 6.87 + (–31.6) ________
7. 9 + 5.68 ________
8. –15.51 + 8.55 ________
9. 36.36 + 54.54 ________

Subtract.

10. 6.23 – 3.62 ________
11. 8.67 – 6.87 ________
12. 28.94 – 9.48 ________
13. 23.57 – 6.84 ________
14. 16.61 – 7.56 ________
15. 32.08 – 12.37 ________
16. 19 – 6.92 ________
17. 42 – 31.89 ________
18. 23 – 21.45 ________
19. 46.2 – 0.27 ________
20. 22 – 18.63 ________
21. 58.9 – 29.58 ________

22. Anna swims the length of the pool in 38.45 seconds and then swims the length of the pool again in 42.38 seconds. What is her total time for 2 lengths of the pool?

____

23. Po has 2 gerbils named Yip and Yap. Yip weighs 3.62 ounces, and Yap weighs 2.79 ounces. How much heavier is Yip than Yap?

____

Name ______________________ Date ______________ Class ______________

LESSON 3-2

Practice C

Adding and Subtracting Decimals

Add or subtract. Estimate to check whether each answer is reasonable.

1. 8.326 + 8.49
2. 13.845 − 10.87
3. 42.35 + 4.368
4. 8.695 − 7.52
5. −5.421 + 18.4
6. −12.54 − 16.951
7. 6.742 − 9.458
8. 8.552 + 14.84
9. −75.25 + (−6.382)
10. 47.68 − 9.654
11. −38.59 − (−7.816)
12. 9.561 + (−16.82)
13. 18.234 + 6.357
14. 123.45 − 12.345
15. 6.456 + (−87.51)
16. 74.832 − 56.842
17. 95.216 − 59.162
18. 82.64 + (−45.846)
19. 27.205 + 19.17 + 8.357
20. 12.63 + 72.029 + 5.9
21. 49.408 + 56.27 − 13.02 − 25.821
22. 9.30 + 17.43 + 38.57 − 25.719
23. A whippet can run a 200-yard course at a rate of 35.5 mi/h. This is 3.85 mi/h slower than the top speed of a greyhound. How fast can a greyhound run?
24. Heike has 2 rocks. If the total weight of both rocks is 54.283 kilograms and the first rock weighs 29.928 kilograms, how much does the second rock weigh?

Name ______________________ Date ______________ Class ______________

LESSON 3-2

Review for Mastery

Adding and Subtracting Decimals

You can use a place-value chart to add decimals.

Add: 3.48 + 2.7

Step 1: Line up the decimal points. Use 0 as a placeholder for hundredths in 2.7.

	Ones	.	Tenths	Hundredths
	3	.	4	8
+	2	.	7	0

Step 2: Add. Place the decimal point in the answer.

	Ones	.	Tenths	Hundredths
	1			
	3	.	4	8
+	2	.	7	0
	6	.	1	8

Step 3: Estimate: 3 + 3 = 6. So, 6.18 is a reasonable answer.

Subtract: 68 – 5.9

Step 1: Line up the decimal points. Use 0 as a placeholder for tenths in 68.

	Tens	Ones	.	Tenths
	6	8	.	0
–		5	.	9

Step 2: Subtract. Place the decimal point in the answer.

	Tens	Ones	.	Tenths
		7		10
	6	~~8~~	.	~~0~~
–		5	.	9
	6	2	.	1

Step 3: Estimate to check: 68 – 6 = 62. So, 62.1 is a reasonable answer.

Add or subtract. Estimate to check your answer.

1. 19.67 + 8.45

	Tens	Ones	.	Tenths	Hundredths
	1	9	.	6	7
+		8	.	4	5
			.		

2. 9.36 – 7.29

	Tens	Ones	.	Tenths	Hundredths
		9	.	3	6
–		7	.	2	9
			.		

3. 25.36 – 9.7

	Tens	Ones	.	Tenths	Hundredths
	2	5	.	3	6
–		9	.	7	
			.		

4. 12.89 + 37.07

	Tens	Ones	.	Tenths	Hundredths
	1	2	.	8	9
+	3	7	.	0	7
			.		

5. 34.71 + 9.19

____

6. 57.06 – 27.38

____

7. 48.3 + (–10.73)

____

Name ______________________ Date ______________ Class ______________

LESSON 3-2

Challenge

Answer Match

Add or subtract. Draw lines to match equivalent answers in each column. Each answer in the middle column matches an answer in both outside columns.

1. −61.244 + 59.014 ________	2. 21.65 + 14.2 + 10.03 ________	3. 41.729 + 17.92 ________
4. 4.018 + 7.052 ________	5. 212.8 − 136.2 ________	6. 42.25 − 13.174 ________
7. −104.02 + 149.9 ________	8. 0.015 − 6.72 ________	9. −0.673 + (−5.707) ________
10. −5.23 − 0.804 + 35.11 ________	11. −70.6 + 48.05 + 33.62 ________	12. 22.015 + 23.865 ________
13. 60.25 − 0.601 ________	14. 65.109 − 5.46 ________	15. 7.17 − 9.4 ________
16. −1.08 + 3.797 − 9.422 ________	17. 10.25 − 12.03 − 0.45 ________	18. 16.02 + 44.48 + 16.1 ________
19. 15.25 + 4.03 + 17.706 ________	20. 130.82 − 101.744 ________	21. −5.61 − 1.095 ________
22. 11.691 − 21.1 + 3.029 ________	23. 45.006 + (−8.02) ________	24. 52.73 − 53.71 + 12.05 ________
25. −33.42 − (−110.02) ________	26. 46.01 + 9.9 − 62.29 ________	27. 6.007 + 74.2 − 43.221 ________

Name ______________________ Date ______________ Class ______________

LESSON 3-2

Problem Solving

Adding and Subtracting Decimals

Write the correct answer.

1. In the Pacific Ocean, the Philippine Trench is 10.05 kilometers deep. In the Atlantic Ocean, the Brazil Basin is 6.12 kilometers deep. How much deeper is the Philippine Trench than the Brazil Basin?

   ____

2. Hawaii's Mauna Kea measures 9.75 kilometers from its base to its peak. The base of Mauna Kea lies 5.55 kilometers below the ocean. What is the height of the part of Mauna Kea that is above sea level?

   ____

3. A team of mountain climbers makes camp 1.48 kilometers above sea level. They climb another 2.91 kilometers to the mountain's peak. How tall is the mountain?

   ____

4. At dawn, the temperature at the summit of a mountain was –8.5 °C. By noon, the temperature had increased 3.6 °C. What was the temperature at noon?

   ____

Choose the letter for the best answer.

This table gives the heights of the tallest mountains on each of the seven continents.

The Seven Summits

Mountain	Country	Height (km)
Mt. Everest	Nepal-Tibet	8.85
Mt. Aconcagua	Argentina	6.96
Mt. McKinley	United States	6.19
Mt. Kilimanjaro	Tanzania	5.90
Mt. Elbrus	Russia	5.64
Vinson Massif	Antarctica	4.90
Puncak Jaya	New Guinea	4.88

5. In 2000, Joby Ogwyn became the youngest person to climb each of the seven summits. How much higher did he climb on Vinson Massif than on Puncak Jaya?

   A 0.02 km
   B 0.12 km
   C 0.18 km
   D 9.78 km

6. Mt. Kilimanjaro, Mt. Elbrus, and Mt. Aconcagua were the first three of the seven summits Joby climbed. What was the total height he climbed?

   F 16.2 km
   G 17.5 km
   H 18.5 km
   J 22 km

7. Mt. Aconcagua is about 1.3 kilometers taller than which mountain?

   A Mt. Everest
   B Mt. Elbrus
   C Mt. McKinley
   D Vinson Massif

Name ______________________ Date ______________ Class ______________

LESSON 3-2

Reading Strategies

Use a Grid

A grid is useful for adding and subtracting decimals.

Add 19.2 + 7.54.

Step 1: Make a grid that has enough squares for each number and one for the decimal point.

Step 2: Place one number in each square of the grid. Carefully line up the decimals.

1	9	.	2	
+	7	.	5	4

Step 3: If there is a square without a number, add a zero as a placeholder.

1	9	.	2	**0**
+	7	.	5	4

Use the grids to answer each question.

1. How do you place the numbers in a grid?

____

2. Why was a zero added to 19.2?

____

3. Write the problem 40.3 – 6.54 in the grid.

4. Did you need to add a zero as a placeholder? If so, where?

____

5. Subtract 6.54 from 40.30 using a grid. What is the answer?

____

Name ______________________ Date ______________ Class ______________

LESSON 3-2

Puzzles, Twisters & Teasers

Up Against the Wall!

Add or subtract to find the solution. Round your solution to the nearest integer. Then answer the riddle.

R 5.37 + 16.45 ________

I 8.89 – 5.91 ________

E 7 + 5.82 ________

L 18.31 – 8.66 ________

N 4.97 – 3.2 ________

M 7.82 + 31.23 ________

O 6 + 9.33 ________

Y 5.98 + 12.99 ________

C 5.02 + 3.21 ________

U 5 – 0.53 ________

H –7.23 + 6.9 ________

A 4.16 – 9.04 ________

T –32.7 + 62.82 ________

What did one wall say to the other wall?

___ ___ ___   ___ ___ ___ ___
3 10 10   39 13 13 30

___ ___ ___   ___ ___   ___ ___ ___
19 15 4   –5 30   30 0 13

___ ___ ___ ___ ___ ___.
8 15 22 2 13 22

Name ______________________ Date ______________ Class ______________

LESSON 3-3

Practice A

Multiplying Decimals

Multiply. Choose the letter for the best answer.

1. 5 • 0.05
   A 25 C 0.25
   B 2.5 D 0.025

2. 9 • 0.7
   F 63 H 0.63
   G 6.3 J 0.063

3. 6 • 0.003
   A 18 C 0.18
   B 1.8 D 0.018

4. 5 • 1.2
   F 60 H 0.6
   G 6 J 0.06

5. 6 • 1.8
   A 10.8 C 0.108
   B 1.08 D 0.0108

6. 8 • 0.02
   F 16 H 0.16
   G 1.6 J 0.016

7. 3 • 8.4
   A 25.2 C 0.252
   B 2.52 D 0.0252

8. 7 • 0.51
   F 357 H 3.57
   G 35.7 J 0.357

Multiply. Estimate to check whether each answer is reasonable.

9. 6.8 • 4 ______________

10. 8.1 • (–2) ______________

11. 9.5 • 5 ______________

12. 3.5 • 7 ______________

13. –6.3 • 6 ______________

14. 9 • 3.7 ______________

15. –6.7 • (–5) ______________

16. 8.8 • (–8) ______________

17. 5.2 • (–4) ______________

18. –3 • 4.1 ______________

19. 1.5 • 1.2 ______________

20 –2.3 • 1.7 ______________

21. Cecile walked 3.7 miles each day for 8 days last month.
    How many miles total did Cecile walk last month?

____

Name ______________________ Date ______________ Class ______________

LESSON 3-3

Practice B

Multiplying Decimals

Multiply.

1. 6 • 0.3 ____________
2. 3 • 0.05 ____________
3. 0.7 • 4 ____________
4. 8 • 6.1 ____________
5. 7.4 • 6 ____________
6. 1.4 • 9 ____________
7. 4.8 • 7 ____________
8. 3 • 8.2 ____________
9. 5.5 • 8 ____________
10. 1.5 • 6 ____________
11. 7.9 • 2 ____________
12. 5 • 6.9 ____________

Multiply. Estimate to check whether each answer is reasonable.

13. 6.3 • 7.8 ____________
14. 9.7 • (–4.7) ____________
15. 6.8 • 0.9 ____________
16. 2.8 • 8.2 ____________
17. –7 • 6.42 ____________
18. 1.9 • 7.22 ____________
19. –5.3 • (–8.4) ____________
20. 7.16 • 0.03 ____________
21. 1.56 • (–7.8) ____________
22. 4.6 • 3.1 ____________
23. 0.62 • 1.45 ____________
24. –5.74 • 1.9 ____________
25. Jordan jogged 4.8 miles each day for 21 days last month. How many miles did she jog last month?

____

Name ______________________ Date ______________ Class ______________

LESSON 3-3

Practice C

Multiplying Decimals

Multiply. Estimate to check whether each answer is reasonable.

1. 6.23 • 5.48 ________
2. 7.82 • 4.18 ________
3. 9.65 • 4.15 ________
4. 2.57 • 3.46 ________
5. 7.84 • 9.23 ________
6. 4.28 • 6.94 ________
7. 42.16 • 8.52 ________
8. 35.84 • 2.57 ________
9. 10.68 • 71.82 ________
10. 6.52 • 3.15 ________
11. 6.78 • (–4.51) ________
12. 7.75 • 7.75 ________
13. 9.49 • 5.25 ________
14. –6.84 • 9.44 ________
15. 10.58 • (–2.54) ________
16. –67.25 • (–5.61) ________
17. 94.21 • (–7.81) ________
18. 5.27 • (–18.94) ________
19. –1.05 • 3.8 • 7.1 ________
20. 0.45 • 1.8 • 0.06 ________
21. 5.7 • 4.52 • (–8.3) ________
22. 0.05 • 1.7 • (–12.62) ________
23. 4.02 • (–9.9) • 23.7 ________
24. –2.2 • 18.45 • (–0.8) ________
25. Phone service costs $24.00 per month plus $0.09 per minute for long distance calls. What is your monthly bill if you make 57 minutes of long distance calls? ________
26. A soccer uniform costs $3.98 for the shirt and $4.49 for the shorts. What is the total cost of 5 uniforms? ________

Name ______________________ Date ______________ Class ______________

LESSON 3-3

Review for Mastery

Multiplying Decimals

To multiply two decimals:

Step 1: Round each number to the nearest integer.

Step 2: Multiply the integers to estimate the product.

Step 3: Multiply the decimals.

Step 4: Place the decimal point in the product to make it closest to the estimate.

Multiply: 2.7 • 4.3

```
   4.3
   2.7
   ---
   301
   860
   ---
 11.61
```

Think:
2.7 rounds to 3.
4.3 rounds to 4.
3 • 4 = 12
Place the decimal point in the product to make it closest to 12.

11.61 is close to 12.

Multiply.

1. 6.7 • 9.1

6.7 rounds to ______________

9.1 rounds to ______________

The product is close to __________.

Product: ______________

2. –3.21 • 8.8

–3.21 rounds to ______________

8.8 rounds to ______________

The product is close to __________.

Product: ______________

3. 4.1 • 0.8

4.1 rounds to ______________

0.8 rounds to ______________

The product is close to __________.

Product: ______________

4. 12.3 • (–2.7)

12.3 rounds to ______________

–2.7 rounds to ______________

The product is close to __________.

Product: ______________

Multiply. Estimate to place the decimal point.

5. 2.06 • 7.9 ______________

6. –4.89 • 0.6 ______________

7. 8.23 • (–4.2) ______________

Name ______________________ Date ______________ Class ______________

LESSON 3-3

Challenge

Decimal Round

Think about numbers that could have resulted in this estimate.

$$0.8 \bullet 3 = 2.4$$

The *least possible factors* that could have been rounded up are 0.75 and 2.5. So, the *least product* is 0.75 • 2.5 = 1.875.

The *greatest possible factors* that could have been rounded down are 0.84 and 3.4. So, the *greatest product* is 0.84 • 3.4 = 2.856.

Find the products of the rounded numbers below. Write the least and greatest factors for each product before rounding. Then find the least and greatest products.

	Rounded Product	Least Product	Greatest Product
1.	18 • 12 =		
2.	3.3 • 4 =		
3.	0.9 • 0.7 =		
4.	1.8 • 2.2 =		
5.	6 • 0.1 =		
6.	4.5 • 0.6 =		
7.	10 • 1.1 =		
8.	0.3 • 15 =		
9.	8.7 • 5.1 =		
10.	1.4 • 1.9 =		

Name ______________________ Date ______________ Class ______________

LESSON 3-3

Problem Solving

Multiplying Decimals

Write the correct answer.

1. A group of 6 adults bought tickets to a play at the community center. The tickets cost $17.50 each. How much did the tickets cost in all?

____

2. Student tickets for the basketball playoffs cost $9.25 each. How much do 9 student tickets cost?

____

3. Movie tickets for senior citizens cost 0.8 times as much as regular adult tickets. Adult tickets cost $7.50. How much do senior citizen tickets cost?

____

____

4. About 30.5 million Americans attend classical music concerts. The average concertgoer attends 2.9 concerts per year. About how many tickets to classical concerts are sold each year?

____

____

Choose the letter for the best answer.

5. The population of Illinois in 2000 was about 1.57 times its population in 1940. If the population of Illinois in 1940 was about 7.9 million, what was its population in 2000, to the nearest tenth of a million?

   A 1.2 million

   B 9.5 million

   C 12.4 million

   D 15.8 million

6. In 2004, the United States had 32.5 million broadband subscribers. This number is expected to increase 1.75 times by 2008. How many broadband subscribers is the United States expected to have in 2008?

   F 24.375 million

   G 34.25 million

   H 56.875 million

   J 568.75 million

7. Nauru and Gibraltar are among the smallest countries in the world. Nauru has about 3.28 times the area that Gibraltar does. Gibraltar is 2.5 square miles. What is the area of Nauru?

   A 0.78 mi^2

   B 0.82 mi^2

   C 5.78 mi^2

   D 8.2 mi^2

8. From 1999 to 2008, the United States Mint is issuing a series of 50 quarters representing each of the 50 states. What is the cost of collecting 10 of each quarter?

   F $100

   G $125

   H $150

   J $250

Name ______________________ Date ______________ Class ______________

LESSON 3-3

Reading Strategies

Compare and Contrast

Decimals are multiplied in much the same way that you multiply whole numbers.

Multiply Whole Numbers	Multiply Decimals
$5 \times 7 = 35$	$0.5 \times 0.7 = 0.35$

Compare multiplying whole numbers to multiplying decimals.

1. What is the same about multiplying whole numbers and decimals?

____

2. What is different about multiplying whole numbers and decimals?

____

____

It is important to place the decimal point correctly in the product.

Steps for Placing the Decimal Point in the Product	**Example:** 1.37 × 0.8
Step 1: Find the product.	1096
Step 2: Count the number of decimal places in each factor.	1.37 ⟶ 2 places 0.8 ⟶ 1 place
Step 3: Find the total number of decimal places in both numbers.	3 places
Step 4: Using the number found in Step 3, move that number of places to the left in the product and place the decimal point.	1.096 (3 2 1)

3. How many decimal places are in 0.63?

____

4. How many decimal places are in 4.231?

____

5. How many decimal places will be in the product of 0.63 × 4.231?

____

Name ________________________ Date ________________ Class ________________

LESSON 3-3

Puzzles, Twisters & Teasers

Turning of the Times!

Multiply the decimals to solve. Round your solution to the nearest integer. Then answer the riddle.

S $9 \cdot 0.36$ ____________

O $8 \cdot 0.6$ ____________

U $6 \cdot 4.9$ ____________

T $2.4 \cdot 3.2$ ____________

I $3.0 \cdot 3.2$ ____________

R $0.6 \cdot 0.05$ ____________

N $3 \cdot 7.01$ ____________

A $-2.6 \cdot 0.4$ ____________

G $8 \cdot 0.77$ ____________

H $1.7 \cdot 12$ ____________

E $2.8 \cdot -1.6$ ____________

W $7 \cdot 0.3$ ____________

When is a car not a car?

___ ___ ___ ___   ___ ___
2 20 –4 21   10 8

___ ___ ___ ___ ___
8 29 0 21 3

___ ___ ___ ___   ___
10 21 8 5   –1

___ ___ ___ ___ ___ ___
6 –1 0 –1 6 –4

Name ______________________ Date ______________ Class ______________

LESSON 3-4

Practice A

Dividing Decimals

Divide. Estimate to check whether your answer is reasonable.

1. $7.5\overline{)15}$

2. $1.2\overline{)72}$

3. $1.5\overline{)45}$

4. $7.5\overline{)-22.5}$

5. $4.8\overline{)16.8}$

6. $-2.7\overline{)11.07}$

Divide. Estimate to check whether your answer is reasonable.

7. $2.8\overline{)14}$

8. $-5.6\overline{)21}$

9. $3.2\overline{)48}$

10. $2.25\overline{)9}$

11. $2.4\overline{)6}$

12. $-1.25\overline{)65}$

13. Jessie used 2.7 gallons of gas to drive her car 72.9 miles. What was her car's gas mileage? ______________

14. Ernesto bicycled 267 miles last week at an average speed of 8.9 mi/h. How many hours did he bicycle? ______________

Name ______________________ Date ______________ Class ______________

LESSON 3-4

Practice B

Dividing Decimals

Divide.

1. 6 ÷ 0.25 ______

2. 78.74 ÷ 12.7 ______

3. 734.8 ÷ −1.67 ______

4. 56.525 ÷ 0.85 ______

5. 44.22 ÷ (−6.7) ______

6. −6.46 ÷ 0.04 ______

Divide. Estimate to check whether your answer is reasonable.

7. 63 ÷ (−4.5) ______

8. 8 ÷ 3.2 ______

9. 87 ÷ 7.25 ______

10. −36 ÷ 1.6 ______

11. 42 ÷ 4.8 ______

12. 90 ÷ 0.36 ______

13. Freddie used 6.75 gallons of gas to drive 155.25 miles. What was his car's gas mileage? ______

14. The members of a book club met at a restaurant for dinner. The total bill was $112.95 and they shared the bill equally. Each person paid $12.55. How many members are there in the book club? ______

Name ______________________ Date ______________ Class ______________

LESSON 3-4

Practice C

Dividing Decimals

Divide. Estimate to check whether each answer is reasonable.

1. $39.984 \div (-5.1)$ ________
2. $355.94 \div 4.81$ ________
3. $-96 \div 1.28$ ________
4. $34.146 \div 6.3$ ________
5. $21.384 \div (-2.4)$ ________
6. $15 \div 0.4$ ________
7. $338.742 \div 0.54$ ________
8. $401.598 \div (-0.81)$ ________
9. $166.17 \div (-0.29)$ ________

Simplify each expression.

10. $40 \cdot (1.8 \div 0.9) \cdot 0.25$ ________
11. $(32.4 \div 6.75) - 2.9 - 4.8$ ________
12. $4.4 + (2.7 - 6.5) \div 0.4 + 0.6$ ________
13. $6.25 \cdot 3.6 \div 1.25 \div 3$ ________
14. $2.3 + (4.2 \cdot 1.32) \cdot 5 + 0.7$ ________
15. $(-7.2 \cdot 0.9) + 3.9 + 6.48$ ________
16. The price of binders at the school store increases by \$0.25 each year. How long will it take for the price to increase by \$3.00? ________

Name ______________________ Date ______________ Class ______________

LESSON 3-4

Review for Mastery

Dividing Decimals

To divide a decimal by a decimal:

Step 1: Make the divisor a whole number by moving the decimal point to the right.

Step 2: Move the decimal point in the dividend the same number of places. Remember to place the decimal in the quotient directly above the decimal point in the dividend.

Step 3: Divide.

Divide: $1.68 \div 0.3 \rightarrow 0.3\overline{)1.68} \rightarrow 3\overline{)16.8}$

```
    5.6
 3)16.8
  -15↓
    18
   -18
```

The divisor, 0.3, has 1 decimal place. Move the decimal point 1 place to the right in <u>both</u> the divisor and the dividend.

Complete.

1. $5.6\overline{)4.48}$

a. How many decimal places are in the divisor? ________

b. How many places do you need to move each decimal point? ________

c. Rewrite the division. ____________

d. Complete the division. What is the quotient? ________

Divide.

2. $5.2\overline{)3.64}$

3. $0.09\overline{)36.45}$

4. $0.59\overline{)0.708}$

Name ______________________ Date ______________ Class ______________

LESSON 3-4

Review for Mastery

Dividing Decimals

Sometimes it is necessary to write zeros in the dividend.

$6 \div 0.25 \longrightarrow 0.25\overline{)6} \longrightarrow 0.25\overline{)6.00} \longrightarrow 25\overline{)600}$

$$\begin{array}{r} 24 \\ 25\overline{)600} \\ \underline{-50\downarrow} \\ 100 \\ \underline{-100} \end{array}$$

The divisor, 0.25, has 2 decimal places. In order to move the decimal point 2 places to the right in the divisor and the dividend, you need to write 2 zeros in the dividend.

Complete.

5. $0.35\overline{)7}$

a. How many decimal places are in the divisor? ________

b. How many places do you need to move each decimal point? ________

c. How many zeros do you need to write in the dividend? ________

d. Complete the division. What is the quotient? ________

Divide.

6. $1.6\overline{)8}$

7. $0.12\overline{)19.2}$

8. $1.25\overline{)48}$

Name ____________________ Date ____________ Class ____________

LESSON 3-4

Challenge

Go For the Gold

A **golden rectangle** is a rectangle with a special relationship between the length and the width. The ancient Greeks believed that the shape of the golden rectangle was especially pleasing to the eye.

Discover the relationship between the length and width that makes a rectangle a golden rectangle.

1. The length of a golden rectangle is 5.26 centimeters and the width is 3.25 centimeters. Use a centimeter ruler to draw the rectangle.

2. The **golden ratio** is a value that characterizes the relationship between the length and width of a golden rectangle. Use the dimensions of the golden rectangle above. Divide the length by the width to find the golden ratio. Round to the nearest hundredth.

____

For each length given below, use the golden ratio to find the width to make a golden rectangle. Round your answers to the nearest hundredth.

3. length = 8 meters ____________
4. length = 4.85 inches ____________
5. length = 6.09 feet ____________
6. length = 12 yards ____________
7. length = 16.5 centimeters ____________
8. length = 24.6 meters ____________
9. length = 55 inches ____________
10. length = 83.2 feet ____________
11. length = 98.2 yards ____________
12. length = 116 centimeters ____________
13. length = 220.5 inches ____________
14. length = 348.7 meters ____________

Name ____________________ Date ______________ Class ______________

LESSON 3-4

Problem Solving

Dividing Decimals

Write the correct answer.

1. Fran has a plank of wood 4.65 meters long. She wants to cut it into pieces 0.85 meters long. How many pieces of wood that length can she cut from the plank?

____

2. Jeremy has a box of 500 nails that weighs 1.35 kilograms. He uses 60 nails to build a birdhouse. How much do the nails in the birdhouse weigh?

____

3. Rhosanda is downloading a file from the Internet. The size of the file is 7.45 MB. The file is downloading at the rate of 0.095 MB per second. How many seconds will it take to download the entire file? Round your answer to the nearest second.

____

4. Sean has a piece of poster board with an area of 476.28 square centimeters. He cuts it into equal-sized squares, each with an area of 39.69 square centimeters. How many squares can he cut from the piece of poster board?

____

Choose the letter for the best answer.

5. Mount McKinley, in Alaska, is about 3.848 miles high. If a mountain climber can climb 0.25 miles per day, about how long, to the nearest day, would it take to climb Mount McKinley?

   A about 5 days
   B about 8 days
   C about 12 days
   D about 15 days

6. In 2000, a production worker in Japan who worked 38.5 hours in a week would have earned an average of $847. What was the hourly wage?

   F $2.20 per hour
   G $3.20 per hour
   H $22.00 per hour
   J $32.00 per hour

7. In the 2000 Summer Olympics, Michael Johnson ran the 400-meter race in 43.84 seconds. To the nearest hundredth, what was his speed in meters per second?

   A 9.12 meters per second
   B 10.96 meters per second
   C 1.09 meters per second
   D 91.24 meters per second

8. In the 2000 Summer Olympics, the United States relay team ran the 1,600-meter relay in 2 minutes, 56.35 seconds. To the nearest hundredth, what was the speed for the relay race in meters per second?

   F 2.83 meters per second
   G 6.24 meters per second
   H 9.07 meters per second
   J 28.39 meters per second

Name ______________ Date ______________ Class ______________

LESSON 3-4

Reading Strategies

Use a Visual Model

John has a piece of lumber 1.5 meters long. He needs to cut it into pieces that are 0.3 meter long. How many pieces can he cut? The number line shows a model of the problem.

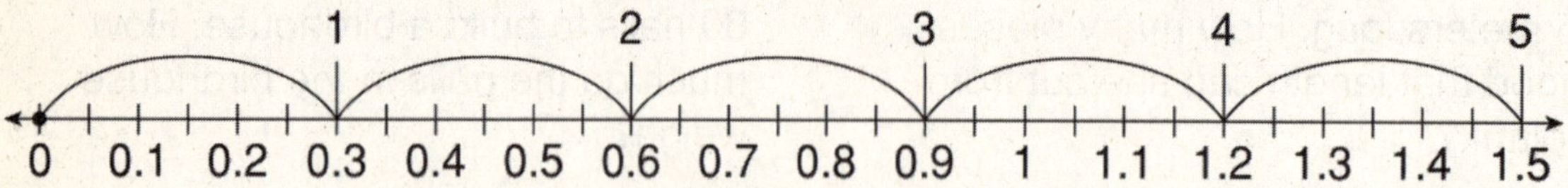

$$1.5 \div 0.3 = 5$$

Sarah has 15 feet of yarn. She needs to cut it into lengths of 3 feet each. How many pieces can she cut? The number line shows a model of the problem.

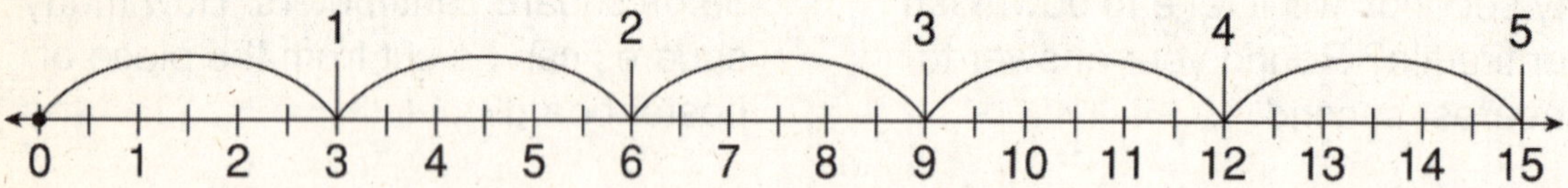

$$15 \div 3 = 5$$

Answer each question.

1. Compare the equations for the number lines above. What is the same about the equations?

2. What is different?

3. Compare the quotients of both problems. What do you notice?

4. How can you change 1.5 to 15?

5. How can you change 0.3 to 3?

6. If you moved the decimal point in **both** the divisor and the dividend, would the quotient change?

Name ______________________ Date ______________ Class ______________

LESSON 3-4

Puzzles, Twisters & Teasers

By the Book!

Match the equations with their answers by writing the letter of the correct answers on the lines. When you are finished, read the letters from left to right and top to bottom as you would any book. If your answers are correct, the letters will spell out the title of a book. Write the book title on the blanks below.

32.1 **E**	0.9 **D**	–15 **N**	–5.3 **O**
5 **O**	6.25 **Y**	–0.5 **H**	120 **O**
0.4 **P**	309 **U**	0.046 **A**	14 **H**
5.6 **R**	–16 **E**	4 **A**	

3.78 ÷ 4.2 = ____	1.06 ÷ –0.2 = ____	7.5 ÷ 1.2 = ____
24 ÷ 0.2 = ____	9.27 ÷ 0.03 = ____	1.12 ÷ 0.08 = ____
–40 ÷ 2.5 = ____	5 ÷ 1.25 = ____	14 ÷ 2.5 = ____
1.15 ÷ 25 = ____	4.12 ÷ 10.3 = ____	–1.6 ÷ 3.2 = ____
12.0 ÷ 2.4 = ____	36 ÷ –2.4 = ____	96.3 ÷ 3 = ____

___ ___   ___ ___ ___   ___ ___ ___ ___

___   ___ ___ ___ ___ ___?

Now circle the name of someone who might have been the author of this book.

By:

Isabel Ringing   Justin Case

M. T. Wallet   Hugo First

Name ______________________ Date ______________ Class ______________

LESSON 3-5

Practice A

Solving Equations Containing Decimals

Solve. Choose the letter for the best answer.

1. $t + 0.7 = 9$
   - A $t = 9.7$
   - B $t = 8.3$
   - C $t = 6.3$
   - D $t = 0.63$

2. $p - 1.6 = 11$
   - F $p = 6.875$
   - G $p = 9.4$
   - H $p = 12.6$
   - J $p = 17.6$

3. $\frac{h}{3} = 1.5$
   - A $h = 0.5$
   - B $h = 4.5$
   - C $h = 9$
   - D $h = 45$

4. $7z = 2.1$
   - F $z = -4.9$
   - G $z = 0.3$
   - H $z = 14.7$
   - J $z = 2.8$

Solve.

5. $x - 5.1 = 4.8$ ______

6. $h + 6.9 = 12.7$ ______

7. $k + 9.2 = -7.6$ ______

8. $g - 4.44 = 2.4$ ______

9. $0.18 + w = 0.75$ ______

10. $m - 3.1 = 9.65$ ______

11. $4.2n = 14.7$ ______

12. $9.7j = 58.2$ ______

13. $56p = -11.76$ ______

14. $43.2 = 2.7y$ ______

15. $64.6 = 6.8x$ ______

16. $40.32 = 12.6m$ ______

17. $\frac{s}{5.4} = 6$ ______

18. $\frac{f}{0.8} = 7$ ______

19. $\frac{d}{4.6} = 0.7$ ______

20. $\frac{c}{0.4} = 1.75$ ______

21. $\frac{h}{6.1} = 12$ ______

22. $\frac{a}{0.35} = 8.4$ ______

23. At the movie theater, a gift certificate for 15 tickets costs $93.75. What is the cost of each ticket? ______

24. Four couples are going to a concert. If each ticket costs $5.90, how much will it cost to buy tickets for all 8 people? ______

Name ____________________ Date ____________ Class ____________

LESSON 3-5

Practice B
Solving Equations Containing Decimals

Solve.

1. $t + 0.77 = 9.3$ ________
2. $p - 1.34 = -11.8$ ________
3. $r + 2.14 = 7.8$ ________
4. $3.65 + e = -1.4$ ________
5. $w - 16.7 = 8.27$ ________
6. $z - 17.2 = 7.13$ ________
7. $p - 67.5 = 24.81$ ________
8. $h + 26.9 = 12.74$ ________
9. $k + 89.2 = -47.62$ ________
10. $x - 0.45 = 5.97$ ________
11. $1.08 + n = 15.72$ ________
12. $y - 6.32 = 0.73$ ________
13. $4.3p = 28.81$ ________
14. $7.7j = 76.23$ ________
15. $3.8g = -104.12$ ________
16. $18.36 = 2.7y$ ________
17. $99.96 = 6.8x$ ________
18. $293.92 = 17.6c$ ________
19. $\frac{e}{7.4} = 6.9$ ________
20. $\frac{f}{12.7} = 15.6$ ________
21. $\frac{d}{9.7} = 20.8$ ________
22. $\frac{w}{-0.2} = 15.4$ ________
23. $\frac{m}{9.8} = 1.7$ ________
24. $\frac{s}{14.35} = -5.2$ ________
25. Jeff paid a flat fee of $269.50 for a year's worth of vet visits for his 4 cats. He made 14 visits during the year. What was the average cost per visit?

____

Name ______________________ Date ______________ Class ______________

LESSON 3-5

Practice C

Solving Equations Containing Decimals

Solve.

1. $t + 4.77 = 9.38$ ______

2. $p - 1.34 = -17.08$ ______

3. $2.55c = 19.89$ ______

4. $3.65 + e = -10.74$ ______

5. $16.17 - w = 8.27$ ______

6. $34.932 = 4.92g$ ______

7. $67.75 - a = 34.81$ ______

8. $h + 26.99 = 11.74$ ______

9. $k + 89.72 = -47.62$ ______

10. $r + 2.14 = 27.84$ ______

11. $6.45j = 36.765$ ______

12. $2.91p = -24.444$ ______

13. $29.406 = 7.54y$ ______

14. $41.096 = 4.67x$ ______

15. $17.52 - z = 4.13$ ______

16. $\frac{a}{7.4} = 6.79$ ______

17. $55.7 + b = 32.119$ ______

18. $\frac{d}{9.7} = 20.78$ ______

19. $0.378 + s = 2.918$ ______

20. $v - 4.025 = 1.9024$ ______

21. $\frac{f}{12.7} = 15.26$ ______

22. $\frac{n}{0.38} = 5.89$ ______

23. $\frac{t}{-1.88} = -6.2$ ______

24. $\frac{w}{0.0056} = -4.7$ ______

25. Sam lives 3.5 times farther from school than Sara does. Sam lives 13.3 miles from school. How far from school does Sara live? ______

26. The Bow-Wow Pet Shop made a profit of $74,239 this year. This was $81,668 more than the profit from last year. What was the profit last year? ______

Name _______________________ Date ______________ Class ______________

LESSON 3-5

Review for Mastery

Solving Equations Containing Decimals

You can solve equations with decimals the same way you solve equations with whole numbers. Remember to always perform the same calculation on both sides of the equation to keep the two sides equal.

- You can use addition to solve a subtraction equation involving decimals.

$$x - 1.45 = 6.7$$
$$x - 1.45 + \mathbf{1.45} = 6.7 + \mathbf{1.45}$$
$$x = 8.15$$

Addition undoes subtraction.

- You can use subtraction to solve an addition equation involving decimals.

$$n + 24.8 = -15.2$$
$$n + 24.8 - \mathbf{24.8} = -15.2 - \mathbf{24.8}$$
$$n = -40$$

Subtraction undoes addition.

Solve.

1. $e + 7.1 = 9.3$

   $e + 7.1 - 7.1 = 9.3 - 7.1$

   $e =$ ____________

2. $x - 1.9 = 5.4$

   $x - 1.9 +$ ______ $= 5.4 +$ ______

   $x =$ ____________

3. $w - 8.3 = -4.12$

   $w - 8.3$ ______ $= -4.12$ ______

   $w =$ ________

4. $b + 5.75 = -6.2$

   $b + 5.75$ ______ $= -6.2$ ______

   $b =$ ________

5. $t + 39.5 = 54.1$

   ____

6. $p - 29.4 = 3.7$

   ____

7. $r - 6.25 = -17.3$

   ____

8. $k + 9.8 = -11.9$

   ____

Name ______________________ Date ______________ Class ______________

LESSON 3-5

Review for Mastery

Solving Equations Containing Decimals (continued)

- You can use division to solve a multiplication equation involving decimals.

$$3.6y = 9$$

$$\frac{3.6y}{3.6} = \frac{9}{3.6}$$

$$y = 2.5$$

Division undoes multiplication.

- You can use multiplication to solve a division equation involving decimals.

$$\frac{a}{4.2} = 18$$

$$\mathbf{4.2} \cdot \frac{a}{4.2} = 18 \cdot \mathbf{4.2}$$

$$a = 75.6$$

Multiplication undoes division.

Solve.

9. $5.7g = 45.6$

$5.7g \div$ ______ $= 45.6 \div$ ______

$g =$ ______

10. $-6f = 8.04$

$-6f$ ______ ______ $= 8.04$ ______ ______

$f =$ ______

11. $\frac{n}{0.14} = 15$

______ $\cdot \frac{n}{0.14} = 15 \cdot$ ______

$n =$ ______

12. $\frac{m}{6.3} = -9.1$

______ ______ $\frac{m}{6.3} = -9.1$ ______ ______

$m =$ ______

13. $8y = 93.6$

______

14. $-3.4c = 20.74$

______

15. $\frac{s}{10.5} = 3.8$

______

16. $\frac{h}{0.4} = -7.2$

______

Name ______________________ Date ______________ Class ______________

LESSON 3-5

Challenge

The Pairs Event

When you solve a system of equations, you solve two different equations at once. The solution to a system of equations is an ordered pair (x, y).

Example: $2.5x = 50$

$x + y = 10.5$

First, solve the first equation for x: $x = 20$.

Then use the solution from the first equation to solve for y in the second equation.

$x = 20$

$$x + y = 10.5$$
$$20 + y = 10.5$$
$$y = -9.5$$

The solution to the system of equations is the ordered pair $(20, -9.5)$.

Solve each system of equations.

1. $x - 5.4 = 7.3$

   $2x + y = 10$

   ____

2. $1.4x = 2.8$

   $\frac{y}{x} = 9.6$

   ____

3. $\frac{x}{3.9} = 0.8$

   $4y = x$

   ____

4. $x + 9.25 = 10.1$

   $y - 3x = 1.7$

   ____

5. $6.4x = 2.56$

   $x = 2y$

   ____

6. $x + 3.02 = 5$

   $\frac{y}{3.5} = x$

   ____

7. $6x = -7.2$

   $y - x = 3.4$

   ____

8. $\frac{x}{0.3} = -9$

   $y + x = 0.6$

   ____

9. $x - 4.7 = -9$

   $\frac{y}{x} = 0.5$

   ____

10. $x + 8.4 = -2$

    $2x + y = 1.1$

    ____

11. $2.6x = 3.64$

    $-3x = y$

    ____

12. $\frac{x}{0.4} = -3.2$

    $x + y = 0.35$

    ____

Name ______________________ Date ______________ Class ______________

LESSON 3-5

Problem Solving

Solving Equations Containing Decimals

Write the correct answer.

1. The diameter of the secondary mirror in NASA's Hubble telescope is 0.3 meter. The primary mirror is 8 times as large. What is the diameter of the primary mirror?

____

2. A cubic centimeter of titanium weighs 4.507 grams. The same volume of gold weighs 19.3 grams. How much more does a cubic centimeter of gold weigh?

____

3. Brianna drives 3.35 miles to work every day. This is 1.75 miles less than Darius drives to work every day. How far does Darius drive to work?

____

4. The weight of an object on Mars is 0.38 times its weight on Earth. How much would a 125-pound person weigh on Mars?

____

Choose the letter for the best answer.

The table shows the orbital velocity of some of the planets.

Planets' Orbital Velocity

Planet	Orbital Velocity (mi/s)
Mercury	29.74
Venus	21.76
Earth	18.5
Mars	14.99
Jupiter	8.12
Saturn	6.02
Uranus	4.23

5. How much greater is the orbital velocity of Mercury than that of Uranus?

   A 33.97 miles per second
   B 25.51 miles per second
   C 7.03 miles per second
   D 4.23 miles per second

6. How many miles does Mercury travel in an hour?

   F 1,784.4 miles
   G 107,064 miles
   H 10,706.4 miles
   J 17,844 miles

7. How many miles does Jupiter travel in a minute?

   A 8.12 miles
   B 81.2 miles
   C 48.72 miles
   D 487.2 miles

8. During the time it takes Saturn to travel 32,508 miles, how much time has elapsed on Earth?

   F 5,400 minutes
   G 1,757.19 minutes
   H 195,698 seconds
   J 90 minutes

Name ______________________ Date ______________ Class ______________

LESSON 3-5

Reading Strategies

Compare and Contrast

Compare the steps for solving equations with whole numbers to the steps for solving equations with decimals.

Solving Equations with Whole Numbers	Example:
Step 1: This is a subtraction problem. Add to get *x* by itself.	$x - 145 = 1{,}720$
Step 2: Add **145** to both sides of the equation.	$x - 145 + \mathbf{145} = 1{,}720 + \mathbf{145}$
Step 3: Solve	$x = 1{,}865$

Solving Equations with Decimals	Example:
Step 1: This is a subtraction problem. Add to get *x* by itself.	$x - 1.45 = 17.2$
Step 2: Add **1.45** to both sides of the equation.	$x - 1.45 + \mathbf{1.45} = 17.2 + \mathbf{1.45}$
Step 3: Solve.	$x = 18.65$

Use the chart to answer the following questions.

1. Compare the steps in solving an equation with whole numbers to the steps for an equation with decimals. What do you notice?

   ____

2. What is different about solving an equation with whole numbers and solving an equation with decimals?

   ____

   ____

Compare solving a multiplication equation with whole numbers to one with decimals: $3y = 702$; $3y = 7.02$. Answer each question.

3. What is the first step in solving both equations?

   ____

4. What operation will you use first in the two equations?

   ____

5. Compare the number you divide by on both sides of the whole number equation to the number you divide by in the decimal equation.

   ____

Name ______________________ Date ______________ Class ______________

LESSON 3-5

Puzzles, Twisters & Teasers

Messy Problems!

Solve the equations. Then use the variables to answer the riddle.

$-1.8 + g = -3.8$ $g =$ ________

$60t = 54$ $t =$ ________

$1.05 = -7n$ $n =$ ________

$h + 0.48 = 1.2$ $h =$ ________

$7.9 = i + 12.7$ $i =$ ________

$e + 0.81 = -6.3$ $e =$ ________

$\frac{l}{0.8} = 4.9$ $l =$ ________

$y - 4.1 = -5$ $y =$ ________

$1.2b = -1.44$ $b =$ ________

$\frac{a}{2.4} = 2.7$ $a =$ ________

$72 = 4.5r$ $r =$ ________

$3.2w = 8$ $w =$ ________

$\frac{d}{-6.4} = 0.85$ $d =$ ________

$s - 9.01 = 12.6$ $s =$ ________

$x + 30.34 = -22.87$ $x =$ ________

Why are basketball players sloppy eaters?

____ ____ ____ ____' ____ ____
0.9 0.72 −7.11 −0.9 16 −7.11

____ ____ ____ ____ ____ ____
6.48 3.92 2.5 6.48 −0.9 21.61

____ ____ ____ ____ ____ ____ ____ ____ ____.
−5.44 16 −4.8 −1.2 −1.2 3.92 −4.8 −0.15 −2

CHAPTER 3

Family Letter

3B Fraction Operations and Applications

Dear Family,

The student will learn how to apply the four basic operations to fractions, starting with addition and subtraction.

When adding and subtracting fractions, the denominators of each fraction have to be the same. If the denominators are already the same, just add the numerators and keep the denominator constant.

When fractions have different denominators, the student will learn how to find a common denominator by either multiplying the denominators or finding the least common multiple (LCM) of the denominators.

Add. Write your answer in simplest form.

$\frac{1}{5} + \frac{1}{6}$

$\frac{1}{5} + \frac{1}{6}$ The LCM of the denominators is 30.

$\frac{6}{30} + \frac{5}{30}$ Write equivalent fractions.

$\frac{11}{30}$ Add.

As the student becomes more proficient in adding and subtracting fractions, the degree of difficulty increases. For example:

Subtract. Write your answer in simplest terms.

$14\frac{1}{2} - 3\frac{3}{4}$

$14\frac{1}{2} \rightarrow 13\frac{6}{4}$ Rename $14\frac{1}{2}$ as $13 + 1\frac{2}{4} = 13 + \frac{4}{4} + \frac{2}{4}$

$-\ 3\frac{3}{4} \rightarrow 3\frac{3}{4}$

$10 + \frac{3}{4}$ Subtract the integers and subtract the fractions.

$10\frac{3}{4}$ Add.

Vocabulary

These are the math words we are learning:

reciprocals two numbers whose product equals 1

CHAPTER 3

Family Letter

3B Fraction Operations and Applications continued

When multiplying fractions the student will need to multiply the numerators **AND** multiply the denominators. If the fractions are mixed numbers, the student will need to first convert the mixed number to an improper fraction.

Multiply. Write each answer in simplest form.

$\mathbf{\frac{1}{5} \cdot 8\frac{4}{5}}$

$\frac{1}{5} \cdot 8\frac{4}{5} = \frac{1}{5} \cdot \frac{44}{5}$	Write the mixed numbers as an improper function.
$= \frac{44}{25}$	Multiply numerators. Multiply denominators.
$= 1\frac{19}{25}$	Simplify.

The student will use the skills from multiplying fractions to divide fractions. He or she will learn to multiply by the **reciprocal** of a fraction when dividing fractions. To find the reciprocal of a number, simply flip or switch the numerator and the denominator of the fraction or whole number. The product of a number and its reciprocal is **ALWAYS** 1.

Divide. Write each answer in simplest form.

$\mathbf{5\frac{1}{2} \div 1\frac{3}{8}}$

$5\frac{1}{2} \div 1\frac{3}{8} = \frac{11}{2} \div \frac{11}{8}$	Write both numbers as improper fractions.
$= \frac{11}{2} \cdot \frac{8}{11}$	Multiply by the reciprocal of $\frac{11}{8}$.
$= \frac{\cancel{11}^{1}}{\cancel{2}_{1}} \cdot \frac{\cancel{8}^{4}}{\cancel{11}_{1}}$	Simplify.
$= 4$	Multiply.

In later chapters, the student will build upon the information learned in this section. Practice adding and subtracting fractions with the student in order to keep these skills sharp. Have him or her begin to recognize the need for fractions in everyday situations.

Sincerely,

Name ______________________ Date ______________ Class ______________

CHAPTER 3

At-Home Practice

3B Fraction Operations and Applications

Estimate each sum or difference.

1. $\frac{3}{8} + 2\frac{5}{6}$

2. $14\frac{4}{7} - 9\frac{8}{9}$

3. $6\frac{2}{3} + 4\frac{1}{4}$

4. Quincy uses $\frac{3}{5}$ of a gallon of milk each day. Estimate how many gallons of milk he needs each week.

Add or subtract. Write each answer in simplest form.

5. $\frac{2}{7} + \frac{3}{7}$

6. $16\frac{3}{4} - 5\frac{1}{2}$

7. $\frac{11}{12} - \frac{2}{3}$

Multiply. Write each answer in simplest form.

8. $2\frac{1}{5} \cdot \left(-\frac{3}{4}\right)$

9. $\frac{2}{3} \cdot \frac{1}{4}$

10. $6 \cdot \frac{5}{7}$

Divide. Write each answer in simplest form.

11. $4\frac{5}{9} \div 2\frac{5}{6}$

12. $\frac{7}{10} \div \left(-1\frac{3}{5}\right)$

13. $9\frac{2}{7} \div 3\frac{5}{7}$

Solve. Write each answer in simplest form.

14. $a + \frac{1}{4} = 3\frac{5}{8}$

15. $\frac{7}{9}x = \frac{1}{6}$

16. $m - \frac{11}{4} = \frac{3}{4}$

Answers: **1.** $3\frac{1}{2}$ **2.** $4\frac{1}{2}$ **3.** 11 **4.** $3\frac{1}{2}$ gallons **5.** $\frac{5}{7}$ **6.** $\frac{45}{4}$ or $11\frac{1}{4}$ **7.** $\frac{1}{4}$ **8.** $-\frac{33}{20}$ or $-1\frac{13}{20}$ **9.** $\frac{1}{6}$ **10.** $\frac{30}{7}$ or $4\frac{2}{7}$ **11.** $\frac{82}{51}$ or $1\frac{31}{51}$ **12.** $-\frac{7}{16}$ **13.** $\frac{5}{2}$ or $2\frac{1}{2}$ **14.** $a = \frac{27}{8}$ or $3\frac{3}{8}$ **15.** $x = \frac{3}{14}$ **16.** $m = \frac{7}{2}$ or $3\frac{1}{2}$

Name ______________________ Date ______________ Class ______________

CHAPTER 3

Family Fun

Fraction Riddles

Directions: Use logical reasoning to answer the questions.

1. When you divide me by $\frac{1}{4}$, the quotient is 25. What number am I?

____

2. When you divide me by $\frac{1}{4}$, the quotient is 42. What number am I?

____

3. When you divide me by $\frac{3}{4}$, the quotient is 3. What number am I?

____

4. When you multiply me by $\frac{3}{4}$, the product is 30. What number am I?

____

5. When you multiply me by $\frac{1}{3}$, the product is 5. What number am I?

____

6. When you multiply me by $\frac{7}{8}$, the product is $5\frac{1}{2}$. What number am I?

____

7. When you multiply me by 2, the product is $\frac{3}{5}$. What number am I?

____

8. When you divide me by $3\frac{1}{6}$, the quotient is $8\frac{1}{4}$. What number am I?

____

9. When you divide me by $1\frac{1}{3}$, the quotient is 3. What number am I?

____

Answers: 1. $6\frac{1}{4}$ **2.** $10\frac{1}{2}$ **3.** $2\frac{1}{4}$ **4.** 40 **5.** 15 **6.** $6\frac{2}{7}$ **7.** $\frac{3}{10}$ **8.** $26\frac{1}{8}$ **9.** 4

Name ______________________ Date ______________ Class ______________

LESSON 3-6

Practice A

Estimating with Fractions

Estimate each sum or difference.

1. $\frac{1}{6} + \frac{5}{8}$

   $0 +$ ____________ $=$ ____________

2. $4\frac{7}{8} - 2\frac{1}{10}$

   ________ $-$ ________ $=$ ________

3. $\frac{4}{5} + \frac{5}{9}$ ____________

4. $\frac{5}{12} + \frac{1}{5}$ ____________

5. $1\frac{7}{8} + 3\frac{5}{6}$ ____________

6. $\frac{7}{8} - \frac{2}{5}$ ____________

7. $\frac{8}{9} - \frac{5}{6}$ ____________

8. $8\frac{3}{5} - 3\frac{9}{10}$ ____________

Estimate each product or quotient.

9. $8\frac{1}{8} \cdot 9\frac{8}{9}$

   $8 \cdot$ ____________ $=$ ____________

10. $39\frac{4}{5} \div 7\frac{9}{10}$

    ________ $\div$ ________ $=$ ________

11. $3\frac{4}{5} \cdot 5\frac{9}{10}$ ____________

12. $5\frac{1}{5} \cdot 3\frac{5}{6}$ ____________

13. $6\frac{3}{8} \cdot 9\frac{1}{8}$ ____________

14. $23\frac{5}{8} \div 6\frac{1}{5}$ ____________

15. $29\frac{7}{9} \div 9\frac{3}{4}$ ____________

16. $9\frac{1}{5} \div 2\frac{7}{8}$ ____________

17. Erin hikes $3\frac{3}{4}$ miles in the morning and $5\frac{1}{8}$ miles in the afternoon. Estimate the number of miles Erin hikes in all.

    ____

18. A trail is $16\frac{3}{8}$ miles long. Max has hiked $4\frac{13}{16}$ miles along the trail so far. Estimate the number of miles Max has left to hike.

    ____

Name ______________________ Date ______________ Class ______________

LESSON 3-6

Practice B

Estimating with Fractions

Estimate each sum or difference.

1. $\frac{5}{11} + \frac{4}{9}$
2. $\frac{16}{13} + \frac{8}{9}$
3. $\frac{9}{10} - \frac{4}{9}$
4. $1\frac{5}{8} - \frac{4}{7}$
5. $3\frac{7}{8} + \left(-\frac{2}{5}\right)$
6. $\frac{8}{9} - \frac{1}{12}$
7. $4\frac{5}{16} + 2\frac{9}{10}$
8. $11\frac{3}{7} - 5\frac{5}{6}$
9. $7\frac{1}{16} + \left(-\frac{11}{12}\right)$

Estimate each product or quotient.

10. $12\frac{2}{5} \div 5\frac{3}{4}$
11. $7\frac{7}{8} \cdot 4\frac{3}{5}$
12. $5\frac{1}{6} \cdot 3\frac{2}{9}$
13. $23\frac{7}{10} \div 4\frac{2}{5}$
14. $17\frac{11}{12} \div 8\frac{5}{9}$
15. $8\frac{7}{12} \cdot 6\frac{9}{10}$
16. $12\frac{3}{8} \cdot 6\frac{1}{6}$
17. $35\frac{2}{3} \div 3\frac{5}{7}$
18. $16\frac{5}{8} \cdot 2\frac{1}{5}$

19. A hallway has a length of $15\frac{3}{4}$ feet and a width of $4\frac{1}{12}$ feet. Estimate the area of the hallway in square feet.

20. A 6-week old puppy weighed $8\frac{7}{16}$ pounds. At 12 weeks of age, the same puppy weighed about $17\frac{3}{8}$ pounds. Estimate how much weight the puppy gained between the ages of 6 weeks and 12 weeks.

Name ______________________ Date ______________ Class ______________

LESSON 3-6

Practice C

Estimating with Fractions

Estimate each sum or difference.

1. $\frac{6}{11} + \frac{6}{7}$

2. $\frac{5}{12} - \frac{2}{5}$

3. $4\frac{4}{9} - 3\frac{1}{8}$

4. $5\frac{4}{7} + \frac{7}{12}$

5. $5\frac{7}{8} + \left(-2\frac{1}{16}\right)$

6. $6\frac{9}{10} - 2\frac{7}{16}$

7. $8\frac{11}{12} + 1\frac{2}{15} - 3\frac{6}{11}$

8. $3\frac{7}{15} + 5\frac{6}{7} - 1\frac{1}{8}$

9. $2\frac{3}{20} - 7\frac{7}{9} + 15\frac{9}{17}$

Estimate each product or quotient.

10. $2\frac{3}{5} \cdot 12\frac{1}{4}$

11. $5\frac{7}{9} \div 2\frac{3}{10}$

12. $20\frac{2}{9} \div \left(-4\frac{5}{7}\right)$

13. $7\frac{5}{16} \cdot \left(-3\frac{5}{8}\right)$

14. $16\frac{7}{9} \cdot 3\frac{9}{20}$

15. $14\frac{11}{15} \div 5\frac{2}{9}$

16. $-6\frac{5}{18} \cdot \left(-7\frac{11}{12}\right)$

17. $47\frac{13}{16} \div 3\frac{2}{11}$

18. $-31\frac{4}{5} \div 7\frac{8}{15}$

19. Carlos weighs $125\frac{1}{4}$ pounds. Hector weighs $17\frac{3}{4}$ pounds less than Carlos. About how much does Hector weigh?

20. A storage closet is $3\frac{3}{4}$ feet wide, $4\frac{1}{6}$ feet long, and $7\frac{11}{12}$ feet high. Estimate the volume of the closet.

Name ______________________ Date ______________ Class ______________

LESSON 3-6

Review for Mastery

Estimating with Fractions

You can estimate sums and differences of fractions and mixed numbers by rounding the fractions or mixed numbers to the nearest $\frac{1}{2}$. Use a number line to help.

Estimate $\frac{3}{8} + \frac{9}{10}$.

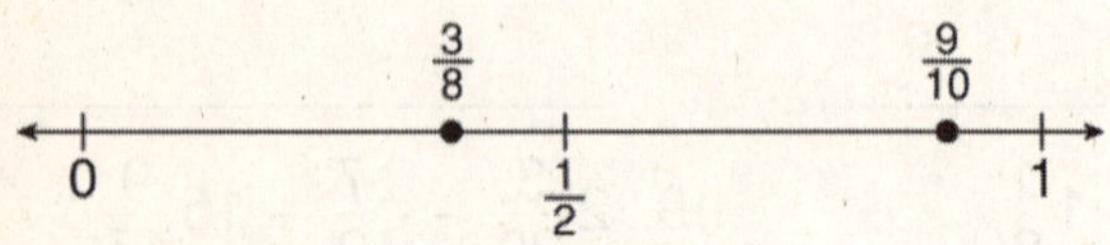

$\frac{3}{8}$ is about $\frac{1}{2}$. $\frac{9}{10}$ is about 1.

So, $\frac{3}{8} + \frac{9}{10}$ is about $\frac{1}{2} + 1$, or $1\frac{1}{2}$.

You can estimate products and quotients of mixed numbers by rounding the mixed numbers to the nearest whole numbers. Use a number line to help.

Estimate $2\frac{1}{4} \cdot 3\frac{5}{8}$.

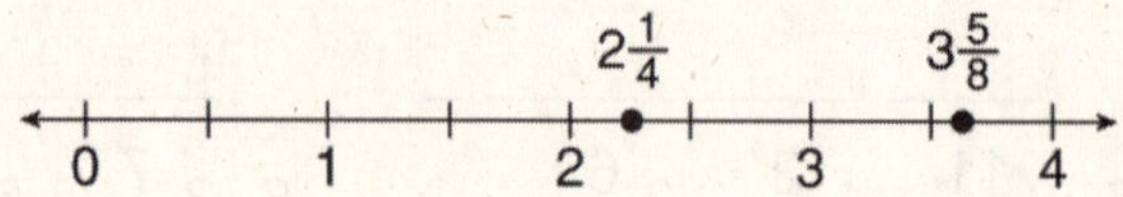

$2\frac{1}{4}$ is about 2. $3\frac{5}{8}$ is about 4.

So, $2\frac{1}{4} \cdot 3\frac{5}{8}$ is about $2 \cdot 4$, or 8.

Estimate.

1. $\frac{7}{8} - \frac{4}{7}$

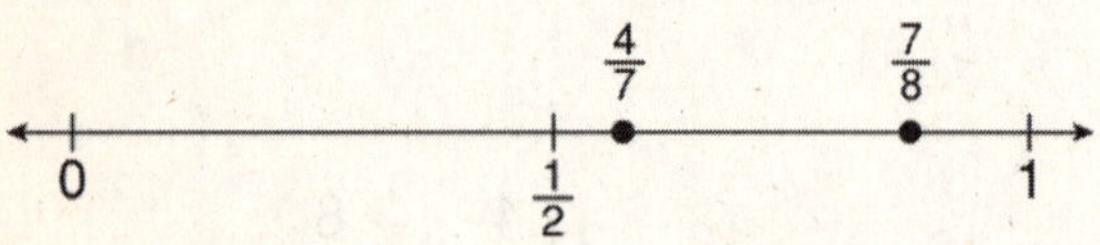

$\frac{7}{8}$ is about ______________

$\frac{4}{7}$ is about ______________.

$\frac{7}{8} - \frac{4}{7}$ is about ________ − ________,

or ________.

2. $4\frac{1}{6} \div 1\frac{5}{9}$

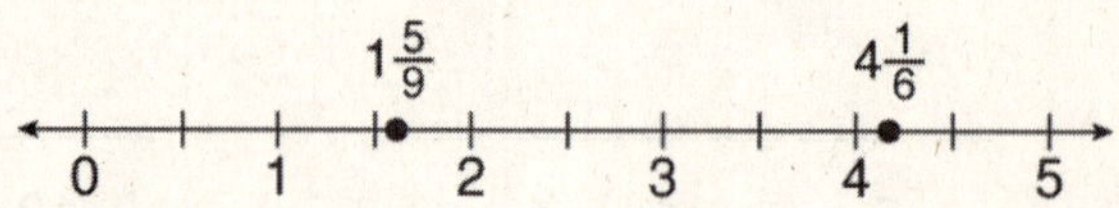

$4\frac{1}{6}$ is about ______________.

$1\frac{5}{9}$ is about ______________.

$4\frac{1}{6} \div 1\frac{5}{9}$ is about ________ ÷ ________,

or ________.

3. $\frac{11}{12} + \frac{2}{5}$ ______________

4. $6\frac{5}{8} - 4\frac{5}{6}$ ______________

5. $5\frac{8}{9} + 3\frac{7}{8}$ ______________

6. $3\frac{3}{8} \cdot 5\frac{7}{9}$ ______________

7. $23\frac{5}{6} \div 7\frac{7}{8}$ ______________

8. $5\frac{1}{4} \cdot 6\frac{2}{5}$ ______________

Name ______________________ Date ______________ Class ______________

LESSON 3-6

Challenge

Don't Underestimate the Answer to a Riddle!

Circle the fractions whose estimated sums, differences products, or quotients equal the given solution. To find the answer to each riddle, write the letters below the circled fractions in order in the blanks.

Why did the student go to school in an airplane?

1. $\frac{7}{8} + \left[1\frac{6}{11} \text{ or } 4\frac{4}{5}\right] - \left[3\frac{7}{8} \text{ or } -\frac{9}{10}\right] \approx 3\frac{1}{2}$

   H E M I

2. $5\frac{11}{12} + \left[6\frac{3}{8} \text{ or } -4\frac{5}{6}\right] \approx 1$

   P G

3. $\left[\frac{6}{11} \text{ or } \frac{6}{7}\right] + \left[5\frac{9}{10} \text{ or } \frac{4}{19}\right] - \left[7\frac{2}{3} \text{ or } -4\frac{16}{19}\right] \approx 5\frac{1}{2}$

   H T Y E S R

ANSWER: She wanted to get a ________________ education!

What kind of hair does the Atlantic Ocean have?

4. $3\frac{1}{10} + \left[\frac{9}{10} \text{ or } 4\frac{5}{12}\right] - \left[-6\frac{1}{20} \text{ or } -6\frac{9}{10}\right] \approx 11$

   W H U A

5. $8\frac{4}{7} - \left[-6\frac{5}{9} \text{ or } -6\frac{1}{8}\right] \approx 15$

   V R

6. $\left[2\frac{1}{6} \text{ or } 1\frac{6}{13}\right] - 4\frac{1}{10} \approx -2$

   Y T

ANSWER: ________________

What can go across the country and still stay in its corner?

7. $\left[4\frac{5}{8} \text{ or } 6\frac{1}{8}\right] \cdot \left[8\frac{3}{10} \text{ or } 12\frac{9}{10}\right] \approx 48$

   T S T R

8. $\left[20\frac{1}{5} \text{ or } 29\frac{7}{8}\right] \div \left[6\frac{3}{5} \text{ or } 4\frac{11}{13}\right] \approx 6$

   U A C M

9. $63\frac{2}{5} \div \left[8\frac{4}{5} \text{ or } 6\frac{7}{9}\right] \approx 7$

   P K

ANSWER: A ________________

Name ______________________ Date ______________ Class ______________

LESSON 3-6

Problem Solving

Estimating with Fractions

Write the correct answer.

1. At the beach, Richard rides the waves on a boogie board that is $3\frac{2}{3}$ feet long. Laura rides a $7\frac{1}{2}$-foot surfboard. Estimate the difference in length of the 2 boards.

____

2. Jorgé had $5\frac{1}{3}$ jugs of apple cider. He used $2\frac{5}{6}$ jugs for a party. About how much apple cider does he have left?

____

3. Sari jogs $2\frac{3}{4}$ miles on Monday, $3\frac{5}{6}$ miles on Wednesday, and $2\frac{1}{3}$ miles on Friday. Estimate her total distance for the week.

____

4. Robert's hand is $2\frac{7}{8}$ inches wide. When Robert uses his hand to estimate the width of his desk, he finds that the desk is about $11\frac{3}{4}$ hands wide. About how many inches wide is the desk?

____

Choose the letter for the best answer.

This table shows the total amount of snow to fall in 5 cities during 2003.

Snowfall During 2003

City	Amount of Snow (in.)
Chicago, IL	$17\frac{2}{5}$
Indianapolis, IN	$44\frac{1}{10}$
Marquette, MI	$191\frac{4}{5}$
Moline, IL	$23\frac{4}{5}$
Providence, RI	$58\frac{9}{10}$

5. About how much snow all together fell in the two cities in Illinois?

A 40 inches C 44 inches
B 41 inches D 61 inches

6. About how much more snow fell in Providence than in Indianapolis?

F 35 inches H 15 inches
G 20 inches J 10 inches

7. Which city had about 11 times as much snow as Chicago?

A Indianapolis C Moline
B Marquette D Providence

8. About how much more snow fell in Indianapolis than in Moline?

F 20 inches H 24 inches
G 22 inches J 30 inches

Name ______________________ Date ______________ Class ______________

LESSON 3-6

Reading Strategies

Analyze Information

You can use some easy rules to help you estimate fractions.

• If the numerator is much smaller than the denominator, round to 0.	**Fractions Close to 0** $\frac{1}{5}$ $\frac{2}{25}$ $\frac{6}{50}$
• If the numerator is about half the value of the denominator, round to $\frac{1}{2}$.	**Fractions Close to $\frac{1}{2}$** $\frac{6}{13}$ $\frac{9}{20}$ $\frac{12}{23}$
• If the numerator and denominator are close to the same value, round to 1.	**Fractions Close to 1** $\frac{7}{8}$ $\frac{23}{25}$ $\frac{57}{60}$

This can help you estimate sums and differences of fractions.

$$\frac{5}{9} - \frac{14}{16} \;\rightarrow\; \frac{1}{2} - 1 = -\frac{1}{2}$$

$$3\frac{3}{8} + 1\frac{8}{9} \;\rightarrow\; 3\frac{1}{2} + 2 = 5\frac{1}{2}$$

Write *close to 0, close to $\frac{1}{2}$, or close to 1* for each fraction.

1. $\frac{13}{25}$ ______________

2. $\frac{9}{75}$ ______________

3. $\frac{7}{9}$ ______________

4. $\frac{14}{16}$ ______________

5. $\frac{3}{20}$ ______________

6. $\frac{1}{9}$ ______________

7. $\frac{7}{15}$ ______________

Name ______________________ Date ______________ Class ______________

LESSON 3-6

Puzzles, Twisters & Teasers

Math Makes Me Crabby!

Estimate the answers. Then solve the riddle.

H $35\frac{1}{3} \div 4\frac{9}{10}$ ____________

T $2\frac{8}{9} + 1\frac{7}{8}$ ____________

S $31\frac{7}{8} \div 3\frac{5}{9}$ ____________

E $2\frac{2}{5} \cdot 2\frac{13}{15}$ ____________

I $\frac{7}{9} - \frac{3}{8}$ ____________

Y $\frac{3}{5} + \frac{6}{7}$ ____________

F $4\frac{5}{7} \cdot 5\frac{1}{3}$ ____________

A $8\frac{3}{4} + -6\frac{1}{5}$ ____________

L $15\frac{1}{7} + 10\frac{8}{9}$ ____________

R $\frac{5}{6} + \frac{2}{12}$ ____________

Why are crabs so greedy?

___ ___ ___ ___   ___ ___ ___
5   7   6 $1\frac{1}{2}$ 3   1   6

___ ___ ___ ___ ___ ___ ___ ___ ___.
8   7   6   26   26   25 $\frac{1}{2}$ 8   7

Name ______________________ Date ______________ Class ______________

LESSON 3-7

Practice A

Adding and Subtracting Fractions

Add or subtract. Write each answer in simplest form.

1. $\frac{1}{2} + \frac{1}{2}$ ________

2. $\frac{2}{5} + \frac{2}{5}$ ________

3. $\frac{1}{4} + \frac{1}{4}$ ________

4. $\frac{1}{8} + \frac{5}{8}$ ________

5. $\frac{5}{6} + \frac{5}{6}$ ________

6. $\frac{7}{20} + \frac{17}{20}$ ________

7. $\frac{7}{12} - \frac{1}{12}$ ________

8. $\frac{9}{16} - \frac{5}{16}$ ________

9. $\frac{3}{10} - \frac{9}{10}$ ________

10. $\frac{7}{8} - \frac{5}{8}$ ________

11. $\frac{4}{9} - \frac{7}{9}$ ________

12. $\frac{7}{15} - \frac{1}{15}$ ________

13. $\frac{1}{3} + \frac{3}{4}$ ________

14. $\frac{1}{4} + \frac{5}{6}$ ________

15. $\frac{5}{8} + \frac{1}{6}$ ________

16. $\frac{1}{4} - \frac{5}{6}$ ________

17. $\frac{3}{4} - \frac{1}{5}$ ________

18. $\frac{9}{10} - \frac{3}{5}$ ________

19. Michael bought $\frac{1}{2}$ pound of Swiss cheese. He used $\frac{1}{4}$ pound for a sandwich. How much was left?

____

Name ______________________ Date ______________ Class ______________

LESSON 3-7

Practice B

Adding and Subtracting Fractions

Add or subtract. Write each answer in simplest form.

1. $\frac{1}{5} + \frac{2}{5}$ ______________

2. $\frac{4}{15} + \frac{8}{15}$ ______________

3. $\frac{7}{12} - \frac{5}{12}$ ______________

4. $\frac{9}{10} - \frac{7}{10}$ ______________

5. $\frac{7}{12} - \frac{11}{12}$ ______________

6. $\frac{2}{7} + \frac{6}{7}$ ______________

7. $\frac{11}{15} + \frac{7}{15}$ ______________

8. $\frac{3}{16} - \frac{1}{16}$ ______________

9. $\frac{8}{21} + \frac{5}{21}$ ______________

10. $\frac{4}{5} - \frac{3}{4}$ ______________

11. $\frac{3}{8} + \frac{1}{2}$ ______________

12. $\frac{2}{5} - \frac{21}{25}$ ______________

13. $\frac{11}{12} + \frac{5}{6}$ ______________

14. $\frac{7}{8} - \frac{5}{12}$ ______________

15. $\frac{9}{10} + \frac{5}{6}$ ______________

16. $\frac{2}{5} - \frac{7}{8}$ ______________

17. $\frac{5}{6} + \frac{11}{15}$ ______________

18. $\frac{3}{4} - \frac{8}{15}$ ______________

19. The school track is $\frac{7}{8}$ mile in length. Sherri ran $\frac{2}{3}$ mile. How much farther does she have to go to get all the way around the track?

____

20. The Millers budget $\frac{1}{2}$ of their income for fixed expenses and $\frac{1}{8}$ of their income for savings. What fraction of their income is left?

____

Name ______________________ Date ______________ Class ______________

LESSON 3-7

Practice C

Adding and Subtracting Fractions

Add or subtract. Write each answer in simplest form.

1. $\frac{7}{15} - \frac{4}{15}$ ______

2. $\frac{7}{18} + \frac{11}{18}$ ______

3. $\frac{6}{7} + \frac{8}{21}$ ______

4. $\frac{2}{5} + \frac{7}{15}$ ______

5. $\frac{5}{12} - \frac{4}{9}$ ______

6. $\frac{7}{30} - \frac{9}{10}$ ______

7. $\frac{8}{9} - \frac{5}{18}$ ______

8. $\frac{7}{25} + \frac{4}{5}$ ______

9. $\frac{3}{8} + \frac{5}{11}$ ______

10. $\frac{1}{8} - \frac{19}{40}$ ______

11. $\frac{5}{8} + \frac{7}{12}$ ______

12. $\frac{5}{6} - \frac{5}{9}$ ______

13. $\frac{7}{8} + \frac{4}{5} + \frac{9}{20}$ ______

14. $\frac{11}{12} - \frac{5}{6} - \frac{2}{3}$ ______

15. $-\frac{2}{3} + \frac{13}{15} - \frac{4}{5}$ ______

16. $-\frac{7}{12} - \frac{1}{5} + \frac{3}{4}$ ______

17. $\frac{7}{9} + \frac{2}{5} - \frac{4}{15}$ ______

18. $\frac{13}{20} - \frac{2}{7} + \frac{11}{35}$ ______

19. In 2001, the population of the United States was about $\frac{2}{7}$ billion people. It is projected that by 2025, there will be over $\frac{1}{3}$ billion people in the United States. How much will the population increase by 2025? ______

20. Benjamin walks $\frac{11}{15}$ mile to work and then $\frac{3}{10}$ mile to the grocery store. How far does he walk? ______

21. About $\frac{2}{5}$ of the population in the United States has blood type A. About $\frac{23}{50}$ have blood type O. What fraction of the population has either blood type A or O? ______

Name ______________________ Date ____________ Class ____________

LESSON 3-7

Review for Mastery

Adding and Subtracting Fractions

To add or subtract fractions with different denominators:

Step 1: Find the least common multiple of the denominators.

Step 2: Write both fractions with the least common multiple (LCM) as the denominator.

Step 3: Add or subtract the numerators, keeping the denominator the same. Write the answer in simplest form.

$\frac{1}{4} + \frac{5}{6}$

The LCM of 4 and 6 is 12.

$\frac{1}{4} = \frac{3}{12}$ and $\frac{5}{6} = \frac{10}{12}$

$\frac{1}{4} + \frac{5}{6} = \frac{3}{12} + \frac{10}{12} = \frac{3+10}{12} = \frac{13}{12} = 1\frac{1}{12}$

$\frac{2}{3} - \frac{1}{2}$

The LCM of 3 and 2 is 6.

$\frac{2}{3} = \frac{4}{6}$ and $\frac{1}{2} = \frac{3}{6}$

$\frac{2}{3} - \frac{1}{2} = \frac{4}{6} - \frac{3}{6} = \frac{4-3}{6} = \frac{1}{6}$

Add or subtract. Write each answer in simplest form.

1. $\frac{3}{5} + \frac{1}{3} = \frac{\quad}{15} + \frac{\quad}{15} = \frac{\quad + \quad}{15} = \frac{\quad}{15}$

2. $\frac{8}{9} - \frac{1}{3} = \frac{8}{9} - \frac{\quad}{9} = \frac{\quad - \quad}{9} = \frac{\quad}{9}$

3. $\frac{2}{5} + \frac{1}{2} = \frac{\quad}{10} + \frac{\quad}{10} = \frac{\quad + \quad}{10} = \frac{\quad}{10}$

4. $\frac{3}{4} - \frac{1}{3} = \frac{\quad}{12} - \frac{\quad}{12} = \frac{\quad - \quad}{12} = \frac{\quad}{\quad}$

5. $\frac{2}{3} + \frac{3}{4} = \frac{\quad}{12} + \frac{\quad}{12} = \frac{\quad + \quad}{12} = \frac{\quad}{\quad} = \frac{\quad}{\quad}$

6. $\frac{1}{4} - \frac{5}{8} = \frac{\quad}{\quad} - \frac{\quad}{\quad} = \frac{\quad - \quad}{\quad} = \frac{\quad}{\quad}$

7. $\frac{1}{4} + \frac{2}{3}$ ____________

8. $\frac{9}{10} - \frac{1}{2}$ ____________

9. $\frac{7}{10} + \frac{2}{5}$ ____________

10. $\frac{11}{12} - \frac{2}{3}$ ____________

11. $\frac{1}{4} - \frac{7}{10}$ ____________

12. $\frac{3}{4} + \frac{4}{5}$ ____________

Name ______________________ Date ______________ Class ______________

LESSON 3-7

Challenge

Fraction Tic-Tac-Toe

Find each sum or difference. Write the answer in simplest terms. Cross off the answers in the tic-tac-toe board to win.

1. $\frac{3}{8} + \frac{1}{2} + \frac{3}{4} =$ ______________

2. $-\frac{6}{7} + \frac{2}{3} + \left(-\frac{11}{21}\right) =$ ______________

3. $\frac{8}{9} + \frac{1}{3} + \left(-\frac{7}{12}\right) =$ ______________

$\frac{1}{2}$	$\frac{1}{7}$	$\frac{23}{36}$
$\frac{2}{12}$	$-\frac{5}{7}$	$\frac{5}{7}$
$1\frac{5}{8}$	$-\frac{1}{6}$	$\frac{7}{8}$

4. $\frac{1}{6} + \left(-\frac{2}{3}\right) - \left(-\frac{5}{12}\right) =$ ______________

5. $-\frac{7}{9} - \frac{5}{6} + \frac{1}{3} =$ ______________

6. $\frac{7}{10} - \left(-\frac{1}{2}\right) + \left(-\frac{3}{5}\right) =$ ______________

$\frac{1}{12}$	1	$-\frac{3}{5}$
$\frac{5}{12}$	$1\frac{5}{18}$	$\frac{4}{5}$
$-\frac{1}{12}$	$-1\frac{5}{18}$	$\frac{3}{5}$

7. $\frac{9}{14} - \left(-\frac{2}{7}\right) - \left(-\frac{1}{2}\right) =$ ______________

8. $-\frac{8}{15} - \frac{1}{3} + \frac{4}{5} =$ ______________

9. $\frac{2}{9} - \frac{2}{5} + \left(-\frac{2}{15}\right) =$ ______________

$-\frac{1}{7}$	$1\frac{3}{7}$	1
$\frac{1}{15}$	$-\frac{1}{15}$	$\frac{1}{7}$
$-\frac{2}{45}$	$-\frac{14}{45}$	$\frac{2}{45}$

Name ______________________ Date ______________ Class ____________

LESSON 3-7

Problem Solving

Adding and Subtracting Fractions

Write the correct answer.

1. During the recycling drive, $\frac{1}{5}$ of the material collected was bottles and $\frac{1}{4}$ was paper. Cardboard boxes made up $\frac{1}{10}$ of the material. How much of the total do these three items represent?

____

2. Decorations for school dances take $\frac{1}{5}$ of the student council's budget. Entertainment takes $\frac{3}{10}$ of the budget. What fraction of the budget is left?

____

3. The school environmental club made a poster to celebrate Earth Day. The poster is $\frac{7}{8}$ yard long and $\frac{2}{3}$ yard wide. What is the difference in the length and width of the poster?

____

4. Three students ran for president of the student council. Eddie received $\frac{1}{5}$ of the votes. Tamara received $\frac{3}{8}$ of the votes. Levi received the rest. Which student won the election?

____

Choose the letter for the best answer.

5. The Reeds budget $\frac{1}{3}$ of their income for rent and $\frac{1}{4}$ for food. How much of their budget is left?

A $\frac{1}{2}$ C $\frac{5}{12}$

B $\frac{3}{4}$ D $\frac{7}{12}$

6. Jasmine's CD collection is $\frac{3}{8}$ jazz, $\frac{1}{4}$ rap, and the rest rock music. What fraction of her CDs is rock music?

F $\frac{1}{4}$ H $\frac{1}{2}$

G $\frac{3}{8}$ J $\frac{5}{8}$

7. Wong has 2 boxes of saltwater taffy. One box contains $\frac{3}{4}$ pound, and the other box contains $\frac{7}{10}$ pound. How much taffy does he have all together?

A $\frac{9}{10}$ pound C $1\frac{1}{2}$ pounds

B $1\frac{9}{20}$ pounds D $1\frac{13}{20}$ pounds

8. In 1992, about $\frac{43}{100}$ people voted for Bill Clinton for President. About $\frac{1}{5}$ voted for Ross Perot and the rest voted for George Bush. About how many voted for George Bush?

F about $\frac{37}{100}$ H about $\frac{57}{100}$

G about $\frac{52}{100}$ J about $\frac{3}{5}$

Name ______________________ Date ______________ Class ______________

LESSON 3-7

Reading Strategies

Use Fraction Strips

It is easy to add and subtract fractions with like, or **common denominators.**

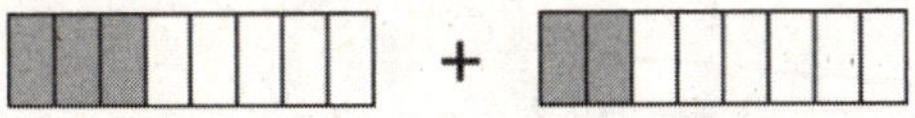

$\frac{3}{8} + \frac{2}{8}$

Use the fraction strips to answer each question.

1. What fractional part of the first fraction strip is shaded? ____________
2. What fractional part of the second fraction strip is shaded? ____________
3. Add the shaded parts of both fraction strips. What fractional part of both is shaded? ____________
4. When you add two fractions with common denominators, does the denominator change? ____________

$\frac{4}{6} - \frac{3}{6}$

Use the fraction strips to answer each question.

5. What fractional part of the first fraction strip is shaded? ____________
6. What fractional part of the second fraction strip is shaded? ____________
7. When you subtract the second fraction strip from the first, what is the answer? ____________
8. When you subtract two fractions, what part of the fraction is subtracted, the numerator or the denominator? ____________

Name ______________________ Date ______________ Class ______________

LESSON 3-7

Puzzles, Twisters & Teasers

Jumping to Conclusions!

Solve the equations. Then answer the riddle.

D $\frac{1}{2} - \frac{3}{4}$ ________

B $\frac{5}{6} - \frac{1}{9}$ ________

E $\frac{1}{2} - \frac{7}{12}$ ________

C $\frac{4}{5} + \frac{6}{7}$ ________

O $\frac{5}{7} + \frac{1}{3}$ ________

A $\frac{3}{4} + \frac{2}{5}$ ________

L $\frac{1}{2} - \frac{2}{7}$ ________

S $\frac{7}{12} + \frac{6}{12}$ ________

I $\frac{1}{5} + \frac{2}{3}$ ________

H $\frac{5}{6} - \frac{1}{6}$ ________

U $\frac{2}{3} + \frac{1}{3}$ ________

T $\frac{21}{24} - \frac{1}{2}$ ________

W $\frac{3}{4} - \frac{11}{12}$ ________

N $\frac{5}{8} + \frac{7}{8}$ ________

If a cat can jump five feet in the air, then why can't it jump through a two-foot high window?

___ ___ ___ ___ ___ ___ ___

$\frac{13}{18}$ $-\frac{1}{12}$ $1\frac{23}{35}$ $1\frac{3}{20}$ 1 $1\frac{1}{12}$ $-\frac{1}{12}$

___ ___ ___   ___ ___ ___ ___ ___ ___

$\frac{3}{8}$ $\frac{2}{3}$ $-\frac{1}{12}$ $-\frac{1}{6}$ $\frac{13}{15}$ $1\frac{1}{2}$ $-\frac{1}{4}$ $1\frac{1}{21}$ $-\frac{1}{6}$

___ ___   ___ ___ ___ ___ ___ ___

$\frac{13}{15}$ $1\frac{1}{12}$ $1\frac{23}{35}$ $\frac{3}{14}$ $1\frac{1}{21}$ $1\frac{1}{12}$ $-\frac{1}{12}$ $-\frac{1}{4}$

Name ______________________ Date ______________ Class ______________

LESSON 3-8

Practice A

Adding and Subtracting Mixed Numbers

Add. Write each answer in simplest form.

1. $5\frac{1}{2} + 2\frac{1}{4}$

2. $3\frac{1}{4} + 4\frac{3}{4}$

3. $2\frac{1}{5} + 1\frac{2}{5}$

4. $1\frac{1}{2} + 3\frac{3}{8}$

5. $2\frac{6}{7} + 5\frac{4}{7}$

6. $3\frac{2}{9} + 3\frac{8}{9}$

7. $2\frac{5}{12} + 3\frac{1}{8}$

8. $2\frac{3}{4} + 5\frac{5}{6}$

9. $\frac{2}{3} + 2\frac{5}{8}$

Subtract. Write each answer in simplest form.

10. $4\frac{3}{4} - 2\frac{1}{4}$

11. $5\frac{5}{6} - 3\frac{1}{6}$

12. $7\frac{2}{3} - 4\frac{1}{3}$

13. $5\frac{1}{4} - 1\frac{3}{4}$

14. $6\frac{1}{3} - 5\frac{2}{3}$

15. $8\frac{1}{6} - 5\frac{5}{6}$

16. $9\frac{5}{6} - 6\frac{1}{4}$

17. $7\frac{3}{4} - 4\frac{5}{6}$

18. $8\frac{3}{8} - 4\frac{3}{4}$

19. Samson bicycled $8\frac{7}{8}$ miles on Friday and $5\frac{1}{4}$ miles on Saturday. How much farther did he ride on Friday?

Name ______________________ Date ______________ Class ______________

LESSON 3-8

Practice B

Adding and Subtracting Mixed Numbers

Add. Write each answer in simplest form.

1. $7\frac{2}{7} + 6\frac{5}{7}$ ______

2. $5\frac{4}{9} + 3\frac{7}{9}$ ______

3. $4\frac{1}{3} + 8\frac{1}{4}$ ______

4. $2\frac{7}{15} + 3\frac{11}{15}$ ______

5. $6\frac{9}{10} + 1\frac{2}{5}$ ______

6. $2\frac{3}{5} + 1\frac{11}{20}$ ______

7. $5\frac{9}{10} + 2\frac{5}{8}$ ______

8. $2\frac{11}{12} + 3\frac{7}{8}$ ______

9. $1\frac{2}{3} + 5\frac{7}{9}$ ______

Subtract. Write each answer in simplest form.

10. $7\frac{7}{9} - 3\frac{5}{9}$ ______

11. $9\frac{7}{10} - 5\frac{3}{10}$ ______

12. $4\frac{13}{15} - 1\frac{7}{15}$ ______

13. $6\frac{2}{3} - 3\frac{3}{5}$ ______

14. $10\frac{3}{4} - 6\frac{1}{3}$ ______

15. $2\frac{3}{10} - 1\frac{7}{8}$ ______

16. $8\frac{7}{12} - 6\frac{1}{3}$ ______

17. $5\frac{7}{8} - 3\frac{9}{10}$ ______

18. $7\frac{6}{7} - 6\frac{3}{4}$ ______

19. Tucker ran $5\frac{3}{8}$ miles on Monday and $3\frac{3}{4}$ miles on Tuesday. How far did he run on both days?

______

Name ________________ Date ________ Class ________

LESSON 3-8

Practice C

Adding and Subtracting Mixed Numbers

Add or subtract. Write each answer in simplest form.

1. $3\frac{11}{21} + 2\frac{8}{21}$ ________
2. $2\frac{5}{26} + 5\frac{11}{13}$ ________
3. $4\frac{2}{11} + 1\frac{6}{22}$ ________
4. $15\frac{16}{19} - 8\frac{7}{19}$ ________
5. $9\frac{13}{25} - 7\frac{2}{5}$ ________
6. $4\frac{11}{20} - 3\frac{3}{10}$ ________
7. $7\frac{3}{5} + 8\frac{1}{8}$ ________
8. $4\frac{2}{3} + 6\frac{3}{7}$ ________
9. $9\frac{5}{12} + 3\frac{3}{8}$ ________
10. $8\frac{5}{6} - 5\frac{3}{5}$ ________
11. $7\frac{7}{12} - 4\frac{11}{15}$ ________
12. $10\frac{2}{3} - 5\frac{13}{18}$ ________
13. $3\frac{2}{5} + 4\frac{1}{3} - 6\frac{7}{10}$ ________
14. $1\frac{1}{4} + 6\frac{7}{8} + 5\frac{5}{16}$ ________
15. $-4\frac{1}{2} - 9\frac{5}{6} - 5\frac{3}{4}$ ________
16. $2\frac{1}{12} - 3\frac{2}{5} + 7\frac{4}{15}$ ________
17. $1\frac{7}{8} + 2\frac{7}{20} - 1\frac{3}{10}$ ________
18. $-6\frac{6}{7} - 3\frac{8}{35} - 1\frac{2}{5}$ ________

Compare. Write <, >, or =.

19. $11\frac{1}{5} - 5\frac{4}{5}$ ____ $7\frac{1}{3} - 1\frac{1}{2}$
20. $7\frac{2}{3} + 4\frac{1}{5}$ ____ $7\frac{5}{6} + 4\frac{2}{7}$
21. $18\frac{1}{2} - 3\frac{3}{4}$ ____ $7\frac{2}{3} + 7\frac{1}{12}$
22. $6\frac{1}{2} - 3\frac{1}{3}$ ____ $5\frac{1}{3} - 2\frac{1}{5}$

23. The men's indoor pole vault record in 1993 was $20\frac{1}{6}$ feet. The women's record in 2001 was $15\frac{5}{12}$ feet. How much higher was the men's record than the women's record? ________

24. Richard set a goal of running 10 miles per week. On Monday he ran $3\frac{3}{5}$ miles. On Friday he ran $3\frac{9}{10}$ miles. How much farther does he still have to run to meet his weekly goal? ________

Name ______________________ Date ______________ Class ______________

LESSON 3-8

Review for Mastery

Adding and Subtracting Mixed Numbers

You can write mixed numbers as improper fractions before adding.

Add: $4\frac{5}{8} + 2\frac{7}{8}$

- Write improper fractions.

$4\frac{5}{8} = \frac{37}{8}$ and $2\frac{7}{8} = \frac{23}{8}$

- The denominators are the same. Add and simplify.

$\frac{37}{8} + \frac{23}{8} = \frac{60}{8} = 7\frac{4}{8} = 7\frac{1}{2}$

Add: $1\frac{2}{5} + 3\frac{3}{4}$

- Write improper fractions.

$1\frac{2}{5} = \frac{7}{5}$ and $3\frac{3}{4} = \frac{15}{4}$

- Find a common denominator. The LCD is 20.

$\frac{7}{5} = \frac{28}{20}$ and $\frac{15}{4} = \frac{75}{20}$

- Add and simplify.

$\frac{28}{20} + \frac{75}{20} = \frac{103}{20} = 5\frac{3}{20}$

Add. Write each answer in simplest form.

1. $4\frac{7}{10} + 2\frac{9}{10} = \frac{\_\_}{10} + \frac{\_\_}{10} = \frac{\_\_}{10} = \frac{\_\_}{10} = \_\_$

2. $2\frac{1}{2} + 1\frac{3}{8} = \frac{\_\_}{2} + \frac{\_\_}{8} = \frac{\_\_}{8} + \frac{\_\_}{8} = \frac{\_\_}{8} = \frac{\_\_}{8}$

3. $3\frac{1}{5} + 2\frac{1}{3} = \frac{\_\_}{5} + \frac{\_\_}{3} = \frac{\_\_}{15} + \frac{\_\_}{15} = \frac{\_\_}{15} = \frac{\_\_}{15}$

4. $1\frac{2}{7} + 5\frac{3}{7}$ ______________

5. $5\frac{3}{8} + 3\frac{7}{8}$ ______________

6. $4\frac{4}{9} + 2\frac{2}{3}$ ______________

7. $2\frac{3}{5} + 3\frac{7}{10}$ ______________

8. $2\frac{3}{4} + 1\frac{5}{6}$ ______________

9. $4\frac{1}{3} + 2\frac{1}{2}$ ______________

Name ______________________ Date ______________ Class ______________

LESSON 3-8

Review for Mastery

Adding and Subtracting Mixed Numbers (continued)

You can write whole numbers as fractions and mixed numbers as improper fractions before subtracting.

$6 - 2\frac{5}{9}$

- Write 6 as a fraction and $2\frac{5}{9}$ as an improper fraction.

$$6 = \frac{6}{1} \text{ and } 2\frac{5}{9} = \frac{23}{9}$$

- Find a common denominator. The LCD is 9.

$$\frac{6}{1} = \frac{54}{9}$$

- Subtract and simplify.

$$\frac{54}{9} - \frac{23}{9} = \frac{31}{9} = 3\frac{4}{9}$$

$4\frac{1}{4} - 1\frac{5}{6}$

- Write improper fractions.

$$4\frac{1}{4} = \frac{17}{4} \text{ and } 1\frac{5}{6} = \frac{11}{6}$$

- Find a common denominator. The LCD is 12.

$$\frac{17}{4} = \frac{51}{12} \text{ and } \frac{11}{6} = \frac{22}{12}$$

- Subtract and simplify.

$$\frac{51}{12} - \frac{22}{12} = \frac{29}{12} = 2\frac{5}{12}$$

Subtract. Write each answer in simplest form.

10. $6\frac{1}{4} - 1\frac{3}{4} = \frac{}{4} - \frac{}{4} = \frac{}{4} = \frac{}{4} = \frac{}{}$

11. $5 - 1\frac{2}{3} = \frac{5}{} - \frac{}{3} = \frac{}{3} - \frac{}{3} = \frac{}{3} = \frac{}{3}$

12. $5\frac{3}{4} - 4\frac{1}{8} = \frac{}{4} - \frac{}{8} = \frac{}{8} - \frac{}{8} = \frac{}{8} = \frac{}{8}$

13. $3\frac{1}{4} - 2\frac{5}{6} = \frac{}{4} - \frac{}{6} = \frac{}{12} - \frac{}{12} = \frac{}{12}$

14. $8\frac{2}{5} - 6\frac{4}{5}$ ______________

15. $5 - 3\frac{4}{7}$ ______________

16. $10 - 5\frac{7}{10}$ ______________

17. $7\frac{1}{2} - 6\frac{3}{10}$ ______________

18. $2\frac{2}{3} - 1\frac{2}{5}$ ______________

19. $5\frac{1}{2} - 3\frac{2}{3}$ ______________

Name ______________________ Date ______________ Class ______________

LESSON 3-8

Challenge

Mixed Number Magic

In a magic square, each row, column, and diagonal has the same sum. Fill in the missing numbers to complete each magic square below.

Magic Square #1

$4\frac{1}{3}$		
$3\frac{1}{4}$		
$8\frac{2}{3}$	$1\frac{1}{12}$	

Magic Square #1 sum: __________

Magic Square #2

$2\frac{1}{6}$	$4\frac{7}{8}$	
	$2\frac{17}{24}$	
		$3\frac{1}{4}$

Magic Square #2 sum: __________

What is the difference between these two magic sums? __________

Magic Square #3

6	$13\frac{1}{2}$	3
	$7\frac{1}{2}$	

Magic Square #3 sum: __________

Magic Square #4

	$3\frac{3}{8}$	$\frac{3}{4}$
	$1\frac{7}{8}$	
	$\frac{3}{8}$	

Magic Square #4 sum: __________

What is the difference between these two magic sums? __________

Name ______________________ Date ______________ Class ______________

LESSON 3-8

Problem Solving

Adding and Subtracting Mixed Numbers

Write the correct answer.

1. A female gray whale is $45\frac{1}{4}$ feet long. A male gray whale is $43\frac{1}{2}$ feet long. How much longer is the female than the male gray whale?

____

2. At birth, a pilot whale is $4\frac{3}{5}$ feet long. A newborn gray whale is $15\frac{1}{4}$ feet long. How much longer is the newborn gray whale?

____

3. A manatee weighs $\frac{1}{2}$ ton. A walrus weighs $1\frac{3}{4}$ tons. A narwhal weighs $1\frac{1}{2}$ tons. What is the total weight of all 3 animals?

____

4. A bottle-nosed dolphin can leap $15\frac{1}{8}$ feet out of the water. The world record high jump for a human is $8\frac{1}{24}$ feet. How much higher can a dolphin leap than a human?

____

Choose the letter for the best answer.

5. At a wildlife park, the killer whale show lasts $\frac{5}{8}$ of an hour. The guided tour of the park takes $2\frac{1}{2}$ hours. How long will it take to do both activities?

A $3\frac{1}{8}$ hours
B $2\frac{5}{8}$ hours
C $3\frac{5}{8}$ hours
D $1\frac{7}{8}$ hours

6. Jeremy walks $3\frac{1}{2}$ miles while visiting a wildlife park. Shawna hikes a $4\frac{3}{8}$-mile long nature trail. How much farther does Shawna walk?

F $\frac{3}{8}$ of a mile
G $\frac{5}{8}$ of a mile
H $\frac{7}{8}$ of a mile
J $1\frac{1}{8}$ of a mile

7. Dog food comes in $5\frac{7}{8}$-pound bags and $12\frac{3}{4}$-pound bags. Find the total weight of 2 small and 1 large bags.

A $22\frac{7}{8}$ pounds
B $23\frac{1}{2}$ pounds
C $24\frac{1}{2}$ pounds
D $25\frac{3}{4}$ pounds

8. A movie lasts $2\frac{1}{6}$ hours. A baseball game lasts $3\frac{2}{3}$ hours. How much longer does the game last?

F $\frac{5}{6}$ hours
G $1\frac{1}{6}$ hours
H $1\frac{1}{3}$ hours
J $1\frac{1}{2}$ hours

Name ______________________ Date ______________ Class ______________

LESSON 3-8

Reading Strategies

Follow a Procedure

To add mixed numbers that have fractions with common denominators, follow these steps.

$2\frac{3}{7} + 1\frac{5}{7}$

Step 1: Add fractions.

$$\begin{array}{r} 2\frac{\mathbf{3}}{\mathbf{7}} \\ +\ 1\frac{\mathbf{5}}{\mathbf{7}} \\ \hline \frac{\mathbf{8}}{\mathbf{7}} \end{array}$$

Step 2: Add integers.

$$\begin{array}{r} \mathbf{2}\frac{3}{7} \\ +\ \mathbf{1}\frac{5}{7} \\ \hline \mathbf{3} \end{array}$$

Step 3: Simplify.

$$\begin{array}{r} 2\frac{3}{7} \\ +\ 1\frac{5}{7} \\ \hline 3\frac{8}{7} = \mathbf{4}\frac{1}{7} \end{array}$$

Use the problem above to answer each question.

1. What part of the mixed numbers should you add first? ______________
2. What part of the mixed numbers should you add next? ______________
3. Do you need to simplify the sum? ______________

When you subtract mixed numbers, you use the same steps as when you add them. However, when the step says add fractions or integers, you will subtract.

Use the problem $5\frac{3}{4} - 2\frac{1}{4}$ to answer each question.

4. What is the first step in solving the problem?

____

5. What is the second step in solving the problem?

____

6. What is the solution?

____

Name ____________________ Date ____________ Class ____________

LESSON 3-8

Puzzles, Twisters & Teasers

Reaching the Right Solutions!

Decide whether or not each equation is correct. Circle the letters above you answers. Then answer the riddle.

1. $70\frac{4}{5} + 1\frac{4}{5} = 72\frac{3}{5}$

R correct A incorrect

2. $7\frac{1}{3} + 8\frac{1}{5} = 16\frac{2}{5}$

Y correct E incorrect

3. $6\frac{1}{4} + 8\frac{3}{4} = 14$

X correct A incorrect

4. $3\frac{4}{5} + 3\frac{2}{5} = 7\frac{1}{5}$

C correct P incorrect

5. $3\frac{1}{2} + 5\frac{1}{4} = 8\frac{3}{4}$

H correct V incorrect

6. $14\frac{3}{5} - 8\frac{1}{2} = 7\frac{1}{5}$

U correct H incorrect

7. $6\frac{3}{4} - 2\frac{3}{4} = 4$

I correct Q incorrect

8. $9\frac{1}{6} - 4\frac{6}{9} = 5\frac{3}{10}$

M correct S incorrect

9. $6\frac{1}{6} + 5\frac{3}{10} = 12\frac{1}{2}$

L correct H incorrect

10. $8 - 2\frac{3}{4} = 5\frac{1}{4}$

A correct K incorrect

11. $3\frac{1}{3} - 2\frac{5}{8} = 6\frac{7}{8}$

J correct N incorrect

12. $1\frac{8}{9} + 4\frac{4}{9} = 5\frac{9}{9}$

G correct D incorrect

13. $6\frac{2}{3} - 5\frac{1}{3} = 1\frac{1}{3}$

S correct F incorrect

Why did the basketball player need long arms?

T O ___ ___ ___ ___ ___

___ ___ ___   ___ ___ ___ ___ ___.

Name ______________________ Date ______________ Class ______________

LESSON 3-9

Practice A

Multiplying Fractions and Mixed Numbers

Multiply. Choose the letter for the best answer.

1. $7 \cdot \frac{1}{8}$

A $\frac{1}{56}$ C $\frac{3}{4}$

B $\frac{7}{15}$ D $\frac{7}{8}$

2. $\frac{2}{5} \cdot \frac{3}{4}$

F $\frac{1}{4}$ H $\frac{3}{10}$

G $\frac{2}{3}$ J $\frac{5}{9}$

3. $4 \cdot 3\frac{3}{5}$

A $2\frac{2}{5}$ C $13\frac{1}{5}$

B 12 D $14\frac{2}{5}$

4. $1\frac{1}{4} \cdot 2\frac{2}{3}$

F $2\frac{1}{6}$ H $3\frac{2}{3}$

G $3\frac{1}{3}$ J $3\frac{11}{12}$

Multiply. Write each answer in simplest form.

5. $4 \cdot \frac{1}{2}$ ______

6. $8 \cdot \frac{1}{4}$ ______

7. $10 \cdot \frac{1}{5}$ ______

8. $\frac{1}{2} \cdot \frac{1}{4}$ ______

9. $\frac{1}{4} \cdot -\frac{2}{3}$ ______

10. $\frac{3}{4} \cdot \frac{2}{3}$ ______

11. $-16 \cdot \frac{3}{4}$ ______

12. $24 \cdot \frac{5}{6}$ ______

13. $32 \cdot \frac{3}{8}$ ______

14. $2\frac{1}{4} \cdot \frac{1}{2}$ ______

15. $3\frac{1}{3} \cdot \frac{3}{5}$ ______

16. $5\frac{1}{3} \cdot \frac{1}{4}$ ______

17. $1\frac{1}{2} \cdot 1\frac{1}{5}$ ______

18. $1\frac{2}{5} \cdot 2\frac{3}{4}$ ______

19. $2\frac{2}{7} \cdot 3\frac{1}{8}$ ______

20. Louis spent 12 hours last week practicing guitar. If $\frac{1}{4}$ of the time was spent practicing chords, how much time did he spend practicing chords? ______

Name ______________________ Date ______________ Class ______________

LESSON 3-9

Practice B

Multiplying Fractions and Mixed Numbers

Multiply. Write each answer in simplest form.

1. $5 \cdot \frac{1}{2}$ ______________

2. $9 \cdot \frac{3}{4}$ ______________

3. $6 \cdot -\frac{2}{5}$ ______________

4. $\frac{9}{15} \cdot \frac{5}{7}$ ______________

5. $\frac{9}{14} \cdot -\frac{7}{9}$ ______________

6. $\frac{7}{12} \cdot \frac{6}{14}$ ______________

7. $-12 \cdot \frac{3}{7}$ ______________

8. $15 \cdot \frac{5}{6}$ ______________

9. $21 \cdot \frac{3}{8}$ ______________

10. $2\frac{1}{3} \cdot \frac{3}{5}$ ______________

11. $3\frac{2}{5} \cdot \frac{1}{2}$ ______________

12. $4\frac{5}{6} \cdot \frac{2}{5}$ ______________

13. $2\frac{2}{5} \cdot \frac{2}{3}$ ______________

14. $3\frac{3}{4} \cdot \frac{2}{5}$ ______________

15. $8\frac{1}{6} \cdot \frac{3}{7}$ ______________

16. $2\frac{1}{3} \cdot 3\frac{3}{8}$ ______________

17. $1\frac{3}{5} \cdot 6\frac{2}{3}$ ______________

18. $2\frac{2}{5} \cdot 4\frac{5}{6}$ ______________

19. Rolf spent 15 hours last week practicing his saxophone. If $\frac{3}{10}$ of the time was spent practicing warm-up routines, how much time did he spend practicing warm-up routines?

____

20. A muffin recipe calls for $\frac{2}{5}$ tablespoon of vanilla extract for 6 muffins. Arthur is making 18 muffins. How much vanilla extract does he need?

____

Name ______________________ Date ______________ Class ______________

LESSON 3-9

Practice C

Multiplying Fractions and Mixed Numbers

Multiply. Write each answer in simplest form.

1. $12 \cdot \frac{1}{7}$ ______

2. $15 \cdot \frac{1}{4}$ ______

3. $-7 \cdot \frac{1}{5}$ ______

4. $\frac{8}{15} \cdot \left(-\frac{3}{4}\right)$ ______

5. $\frac{6}{15} \cdot \frac{7}{18}$ ______

6. $\frac{9}{11} \cdot \frac{22}{27}$ ______

7. $\frac{2}{9} \cdot \frac{7}{10}$ ______

8. $\frac{2}{15} \cdot \frac{5}{12}$ ______

9. $\frac{7}{12} \cdot \frac{3}{7} \cdot \frac{1}{2}$ ______

10. $3\frac{1}{2} \cdot 2\frac{5}{7}$ ______

11. $3\frac{3}{8} \cdot 4\frac{2}{5}$ ______

12. $-4\frac{1}{8} \cdot \frac{2}{9}$ ______

13. $2\frac{2}{11} \cdot \frac{2}{3}$ ______

14. $3\frac{3}{5} \cdot \frac{2}{9} \cdot \frac{2}{3}$ ______

15. $5\frac{2}{5} \cdot 3\frac{1}{9}$ ______

Complete each multiplication sentence.

16. $\frac{?}{5} \cdot \frac{3}{10} = \frac{3}{25}$ ______

17. $\frac{1}{4} \cdot \frac{?}{7} = \frac{3}{14}$ ______

18. $\frac{2}{3} \cdot \frac{?}{8} = \frac{5}{12}$ ______

19. $\frac{?}{3} \cdot \frac{3}{8} = \frac{1}{4}$ ______

20. $\frac{4}{5} \cdot \frac{?}{4} = \frac{3}{5}$ ______

21. $\frac{3}{4} \cdot \frac{?}{7} = \frac{3}{7}$ ______

22. A hippopotamus lives about $2\frac{4}{7}$ times as long as a tiger. A tiger lives an average of 16 years. How long does the average hippopotamus live? ______

23. A pie crust recipe calls for $\frac{4}{5}$ teaspoon of baking powder. Alice is making 13 pies. How much baking powder does she need? ______

Name ______________________ Date ______________ Class ______________

LESSON 3-9

Review for Mastery

Multiplying Fractions and Mixed Numbers

To multiply fractions and mixed numbers:

Step 1: Write any mixed numbers as improper fractions.

Step 2: Multiply the numerators.

Step 3: Multiply the denominators.

Step 4: Write the answer in simplest form.

Remember, positive times negative equals negative.

Multiply: $\frac{4}{9} \cdot \frac{3}{8}$

$$\frac{4}{9} \cdot \frac{3}{8} = \frac{4 \cdot 3}{9 \cdot 8} = \frac{12}{72} = \frac{1}{6}$$

Divide numerator and denominator by 12, the GCF.

Multiply: $6\frac{1}{4} \cdot \left(-1\frac{4}{5}\right)$

$$6\frac{1}{4} \cdot \left(-1\frac{4}{5}\right) = \frac{25}{4} \cdot \left(\frac{-9}{5}\right) = \frac{25 \cdot (-9)}{4 \cdot 5} = \frac{-225}{20} = -11\frac{1}{4}$$

Multiply. Write each answer in simplest form.

1. $6 \cdot \frac{1}{9} = \frac{6 \cdot 1}{9} = \frac{}{} = \frac{}{}$

2. $-\frac{4}{5} \cdot \frac{5}{7} = -\frac{4 \cdot }{5 \cdot } = -\frac{}{} = -\frac{}{}$

3. $3\frac{1}{3} \cdot 9 = \frac{10}{3} \cdot 9 = \frac{10 \cdot }{} = \frac{}{} =$

4. $\frac{3}{10} \cdot 2\frac{1}{2} = \frac{3}{10} \cdot \frac{5}{2} = \frac{ \cdot }{ \cdot } = \frac{}{} = \frac{}{}$

5. $\frac{2}{7} \cdot \frac{7}{8}$ ______________

6. $-\frac{5}{9} \cdot \frac{3}{4}$ ______________

7. $\frac{9}{10} \cdot \left(-\frac{2}{3}\right)$ ______________

8. $2\frac{5}{8} \cdot \frac{2}{3}$ ______________

9. $\frac{1}{2} \cdot 4\frac{1}{4}$ ______________

10. $-\frac{2}{3} \cdot 1\frac{3}{4}$ ______________

11. $5\frac{1}{5} \cdot \left(-1\frac{2}{3}\right)$ ______________

12. $4\frac{1}{2} \cdot 1\frac{1}{9}$ ______________

13. $-2\frac{3}{4} \cdot \left(-1\frac{1}{3}\right)$ ______________

Name ______________________ Date ______________ Class ______________

LESSON 3-9

Challenge

Weigh Out in Space

How much do you think you would weigh on Mars? You can calculate anyone's weight on any of the planets by multiplying his or her weight times the planet's gravitational pull. Use the table to calculate what each animal listed below would weigh on different planets. Write your answers in simplest form.

Gravitational Pull of the Planets

Planet	Mercury	Venus	Mars	Jupiter	Saturn	Uranus	Neptune
Gravitational Pull	$\frac{19}{50}$	$\frac{9}{10}$	$\frac{19}{50}$	$2\frac{1}{2}$	$\frac{19}{20}$	$\frac{4}{5}$	$1\frac{1}{5}$

	Animal	**Weight on Earth**	**Planet**	**Weight on Other Planet**
1.	Mouse	$\frac{1}{16}$ pound	Neptune	
2.	Cat	$5\frac{1}{4}$ pounds	Mercury	
			Jupiter	
3.	Dog	$50\frac{1}{2}$ pounds	Saturn	
			Neptune	
4.	Gorilla	200 pounds	Mars	
			Uranus	
5.	Cow	1,000 pounds	Venus	
			Jupiter	
6.	Small parrot	$\frac{7}{8}$ pound	Mercury	
			Neptune	
7.	Macaw	$3\frac{1}{2}$ pounds	Neptune	
8.	Elephant	10,700 pounds	Uranus	

Name ______________________ Date ______________ Class ______________

LESSON 3-9

Problem Solving

Multiplying Fractions and Mixed Numbers

Write the correct answer.

1. Ariel's English homework is to read 24 pages. She reads $\frac{1}{8}$ of the assignment on the bus ride home. How many pages does she read on the bus?

____

2. When a group of 40 friends goes to the movies at a multiplex, $\frac{1}{5}$ of the group decides to watch a science fiction movie. How many of the group see the science fiction movie?

____

3. As of 1990, the American Indian population was about 2,000,000. About $\frac{1}{5}$ were Cherokee. About how many members of the Cherokee tribe were there in 1990?

____

4. Ron spends 3 hours painting a picture. Ashley spends $2\frac{2}{3}$ as long creating a sculpture. How long does Ashley work on her sculpture?

____

Choose the letter for the best answer.

5. One cup of dry dog food weighs $1\frac{4}{5}$ ounces. A K9 dog eats $6\frac{1}{3}$ cups of food a day. How many ounces of food does the dog eat each day?

A $4\frac{8}{15}$ ounces

B $8\frac{2}{15}$ ounces

C $11\frac{2}{5}$ ounces

D $3\frac{14}{27}$ ounces

6. Max has enough chicken to make $8\frac{1}{2}$ servings of salad. He needs $\frac{1}{8}$ pound of chicken per serving. How much chicken does he have?

F 8 pounds

G $1\frac{1}{16}$ pounds

H $8\frac{5}{8}$ pounds

J $2\frac{1}{2}$ pounds

7. A meteorite found in Willamette, Oregon, weighed $\frac{7}{10}$ as much as one found in Armanti, Western Mongolia. The meteorite found in Armanti weighed 22 tons. How much did the one in Oregon weigh?

A $31\frac{1}{5}$ tons

B $21\frac{3}{10}$ tons

C $22\frac{7}{10}$ tons

D $15\frac{2}{5}$ tons

8. Nicole takes part in a $12\frac{1}{2}$-mile walk-a-thon to raise money for charity. She stops $\frac{1}{2}$ of the way to rest. How much farther must Nicole walk to finish the walk-a-thon?

F $6\frac{1}{4}$ miles

G $6\frac{1}{2}$ miles

H 12 miles

J 25 miles

Name ________________ Date ________________ Class ________________

LESSON 3-9

Reading Strategies

Use Fraction Strips

You can write a multiplication problem as a repeated addition problem.

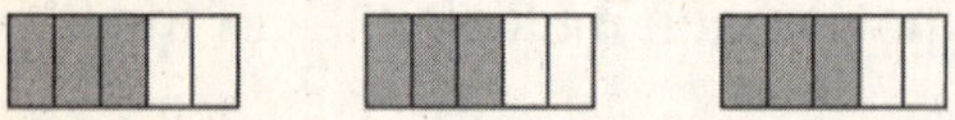

$\frac{3}{5}$ + $\frac{3}{5}$ + $\frac{3}{5}$

$\frac{3}{5} + \frac{3}{5} + \frac{3}{5}$ ← Repeated addition

Use the fraction strips above to answer questions 1–4.

1. What fractional part of the fraction strips is shaded? ________
2. How many fraction strips are there? ________
3. Count the number of fractional parts that are shaded in all. How many are there? ________
4. How can you find the answer to the problem above using addition?

____

You can also find the answer to the above problem using multiplication. $3 \times \frac{3}{5} = \frac{9}{5}$

Use the fraction strips below to answer questions 5–7.

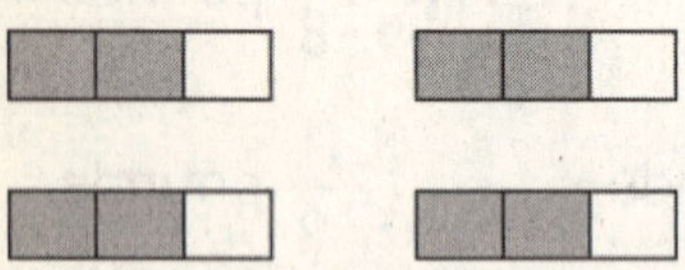

5. What fractional part of each fraction strip is shaded? ________
6. How many of these fraction strips are there? ________
7. Write a multiplication equation for this picture. ________

Name ______________________ Date ______________ Class ______________

LESSON 3-9

Puzzles, Twisters & Teasers

Wooden It Be Nice?

Decide whether or not each equation is correct. Circle the letter above your answer. Use the letters you circled to solve the riddle.

1. $2 \cdot \frac{3}{8} = 0.75$

   I — correct L — incorrect

2. $-15 \cdot \frac{2}{3} = -12$

   M — correct T — incorrect

3. $\frac{1}{4} \cdot \frac{4}{5} = \frac{1}{5}$

   W — correct N — incorrect

4. $\frac{1}{3} \cdot 4\frac{1}{2} = 1\frac{1}{2}$

   O — correct P — incorrect

5. $3\frac{3}{5} \cdot 1\frac{1}{12} = 3\frac{7}{10}$

   Q — correct O — incorrect

6. $\frac{1}{2} \cdot \frac{1}{3} \cdot \frac{1}{4} = \frac{1}{16}$

   R — correct D — incorrect

7. $\frac{1}{6} \cdot 12\frac{1}{2} = 24$

   S — correct E — incorrect

8. $2\frac{1}{4} \cdot 1\frac{1}{2} = 3\frac{3}{8}$

   N — correct J — incorrect

9. $2 \cdot 5\frac{2}{3} = 10\frac{1}{3}$

   H — correct G — incorrect

10. $75 \cdot 1\frac{1}{4} = 93.75$

    O — correct K — incorrect

What happened to the wooden plane with the wooden wheels and wooden engine?

___ ___

___ ___ ___ ___ ___ ___

___ ___

Name ____________ Date ____________ Class ____________

LESSON 3-10

Practice A

Dividing Fractions and Mixed Numbers

Divide. Write each answer in simplest form.

1. $5 \div \frac{1}{2}$ ____________

2. $9 \div \frac{1}{3}$ ____________

3. $6 \div \frac{1}{4}$ ____________

4. $3 \div \frac{3}{4}$ ____________

5. $10 \div \frac{5}{6}$ ____________

6. $6 \div \frac{3}{8}$ ____________

Divide. Find each quotient in the box.

$\frac{1}{5}$	$\left(-\frac{1}{4}\right)$	$\frac{1}{2}$	$\left(-\frac{6}{11}\right)$	$\frac{5}{7}$	$\frac{7}{8}$	1	$1\frac{1}{2}$	2	$2\frac{6}{7}$	3	4	$5\frac{1}{3}$	$7\frac{1}{2}$

7. $\frac{9}{5} \div \frac{3}{5}$ ____________

8. $\frac{6}{7} \div \frac{3}{7}$ ____________

9. $\frac{1}{6} \div \frac{5}{6}$ ____________

10. $\frac{1}{3} \div \frac{2}{3}$ ____________

11. $\frac{3}{4} \div \frac{1}{2}$ ____________

12. $\frac{1}{6} \div \left(-\frac{2}{3}\right)$ ____________

13. $2\frac{2}{3} \div \frac{1}{2}$ ____________

14. $1\frac{1}{4} \div \frac{1}{6}$ ____________

15. $2\frac{1}{2} \div \frac{7}{8}$ ____________

16. $2\frac{1}{2} \div 3\frac{1}{2}$ ____________

17. $1\frac{1}{6} \div 1\frac{1}{3}$ ____________

18. $\left(-1\frac{1}{5}\right) \div 2\frac{1}{5}$ ____________

19. A jug holds 12 pints of juice. How many $1\frac{1}{2}$ pint bottles can be filled with that much juice?

____

Name ____________________ Date ____________ Class ____________

LESSON 3-10

Practice B

Dividing Fractions and Mixed Numbers

Divide. Write each answer in simplest form.

1. $4 \div \frac{1}{2}$

2. $\frac{1}{5} \div \frac{1}{4}$

3. $\frac{1}{3} \div \frac{3}{5}$

4. $\frac{8}{9} \div \frac{2}{3}$

5. $-\frac{3}{8} \div \frac{3}{4}$

6. $\frac{7}{10} \div \frac{3}{5}$

7. $\frac{5}{12} \div \frac{2}{5}$

8. $\frac{3}{4} \div \frac{4}{9}$

9. $\frac{7}{12} \div \frac{3}{4}$

10. $-4\frac{1}{6} \div \frac{1}{3}$

11. $3\frac{1}{4} \div \frac{2}{5}$

12. $6\frac{1}{9} \div \frac{1}{6}$

13. $2\frac{1}{4} \div 1\frac{3}{4}$

14. $3\frac{3}{4} \div 2\frac{5}{6}$

15. $5\frac{1}{3} \div -1\frac{4}{5}$

16. $2\frac{1}{2} \div 2\frac{1}{3}$

17. $-1\frac{3}{4} \div 1\frac{1}{4}$

18. $7\frac{2}{3} \div 1\frac{1}{5}$

19. Burger Barn has $46\frac{2}{3}$ pounds of ground beef. How many $\frac{1}{3}$-pound burgers can be made using all the ground beef?

20. Roberto needs some roofing tiles to be cut from a large tile. How many tiles that are each $14\frac{3}{8}$ inches in length can he cut from a larger piece of tile that is $100\frac{5}{8}$ inches long?

Name ______________________ Date ______________ Class ______________

LESSON 3-10

Practice C

Dividing Fractions and Mixed Numbers

Divide. Write each answer in simplest form.

1. $17 \div \frac{4}{9}$ ________

2. $23 \div \frac{5}{7}$ ________

3. $39 \div \frac{7}{11}$ ________

4. $-\frac{11}{15} \div \frac{7}{9}$ ________

5. $\frac{9}{14} \div \frac{11}{21}$ ________

6. $\frac{15}{19} \div \frac{3}{8}$ ________

7. $\frac{3}{25} \div \frac{11}{15}$ ________

8. $\frac{5}{39} \div -\frac{5}{13}$ ________

9. $\frac{7}{17} \div \frac{14}{17}$ ________

10. $5\frac{3}{7} \div \frac{4}{9}$ ________

11. $7\frac{8}{9} \div \frac{5}{11}$ ________

12. $-16\frac{2}{11} \div \frac{5}{6}$ ________

13. $8\frac{5}{7} \div 3\frac{6}{7}$ ________

14. $9\frac{3}{8} \div 2\frac{7}{8}$ ________

15. $5\frac{3}{5} \div 1\frac{3}{8}$ ________

16. $-\frac{2}{9} \div \frac{7}{8} \div \frac{1}{3}$ ________

17. $\frac{3}{5} \div 4\frac{1}{4} \div \frac{2}{5}$ ________

18. $3\frac{2}{3} \div \frac{2}{7} \div \frac{7}{9}$ ________

19. Jorge can mow 1 lawn in $\frac{3}{4}$ hour. How many lawns can he mow in $3\frac{3}{4}$ hours?

____

20. How many $2\frac{1}{2}$-foot shelves can you cut from a $17\frac{1}{2}$-foot board?

____

Name ______________________ Date ______________ Class ______________

LESSON 3-10

Review for Mastery

Dividing Fractions and Mixed Numbers

Dividing fractions and mixed numbers is very much like multiplying fractions and mixed numbers. Just follow these steps:

Step 1: Write any mixed numbers as improper fractions.
Step 2: Invert the divisor.
Step 3: Multiply and write the quotient in simplest form.

Divide: $1\frac{1}{8} \div \frac{1}{3}$

Step 1: $1\frac{1}{8} \div \frac{1}{3} = \frac{9}{8} \div \frac{1}{3}$

Step 2: $\frac{9}{8} \div \frac{1}{3} = \frac{9}{8} \cdot \frac{3}{1}$

Step 3: $\frac{9}{8} \cdot \frac{3}{1} = \frac{27}{8} = 3\frac{3}{8}$

Divide: $1\frac{1}{4} \div 3\frac{1}{3}$

Step 1: $1\frac{1}{4} \div 3\frac{1}{3} = \frac{5}{4} \div \frac{10}{3}$

Step 2: $\frac{5}{4} \div \frac{10}{3} = \frac{5}{4} \cdot \frac{3}{10}$

Step 3: $\frac{5}{4} \cdot \frac{3}{10} = \frac{15}{40} = \frac{3}{8}$

Divide. Write each answer in simplest form.

1. $\frac{4}{5} \div \frac{1}{2} = \frac{4}{5} \cdot$ ____ = ____ = ____

2. $\frac{5}{8} \div \frac{5}{6} = \frac{5}{8} \cdot$ ____ = ____ = ____

3. $2\frac{1}{2} \div 1\frac{3}{4} = \frac{\_}{2} \div \frac{\_}{4} = \frac{\_}{2} \cdot$ ____

   = ____ = ____ = ____

4. $2\frac{2}{3} \div 1\frac{1}{5} = \frac{\_}{3} \div \frac{\_}{5} = \frac{\_}{3} \cdot$ ____

   = ____ = ____ = ____

5. $\frac{3}{5} \div \frac{3}{10}$ ____________

6. $\frac{7}{8} \div \frac{1}{3}$ ____________

7. $\frac{5}{12} \div \frac{1}{2}$ ____________

8. $4\frac{1}{3} \div 1\frac{1}{9}$ ____________

9. $2\frac{1}{3} \div 1\frac{3}{4}$ ____________

10. $5\frac{5}{8} \div 2\frac{1}{2}$ ____________

Name ______________________ Date ______________ Class ______________

LESSON 3-10

Challenge

Animal Senior Citizens

The value of each division expression will show the average life expectancy of an animal. Complete the chart. Then use the information to fill in the blanks below.

	Animal	Division Expression	Life Expectancy (yr)
1.	Black bear	$2\frac{1}{2} \div \frac{5}{36}$	
2.	Chimpanzee	$40\frac{2}{5} \div 2\frac{1}{50}$	
3.	Dog	$15\frac{1}{2} \div 1\frac{7}{24}$	
4.	Hippopotamus	$4\frac{3}{8} \div \frac{7}{64}$	
5.	Mouse	$\frac{5}{6} \div \frac{5}{18}$	
6.	Guinea Pig	$\frac{2}{3} \div \left(-\frac{3}{5}\right) \div \left(-\frac{5}{18}\right)$	
7.	Tiger	$-2\frac{1}{4} \div 3\frac{1}{2} \div \left(-\frac{9}{224}\right)$	
8.	Kangaroo	$\left(\frac{3}{4} + \frac{7}{8}\right) \div \frac{13}{56}$	

9. If you divide the average life span of a guinea pig by ________, you get the average life span of a hippopotamus.

10. If you divide the average life span of a kangaroo by $\frac{7}{20}$, you find the average life span of a horse, which is ____________.

11. A ____________ has an average life span that is 37 years less than a ____________, which has an average life span that is $5\frac{1}{2} \div 2\frac{3}{4}$ years longer than that of an African elephant, which is ____________.

12. (Average life expectancy of a dog) ÷ $\frac{3}{8}$ ÷ $6\frac{2}{5}$ = (average life expectancy of a rabbit) = ____________.

Name ______________________ Date ____________ Class ____________

LESSON 3-10

Problem Solving

Dividing Fractions and Mixed Numbers

Write the correct answer.

1. The Wheeling Bridge in West Virginia is about $307\frac{4}{5}$ meters long. If you walk with a stride of about $\frac{3}{10}$ meter, how many steps would it take you to cross this suspension bridge?

   ____

2. The Flathead Rail Tunnel in Montana is about $7\frac{3}{4}$ miles long. If a train travels through the tunnel at a speed of about $1\frac{1}{2}$ miles per minute, how long will it take to pass from one end of the tunnel to the other?

   ____

3. A hiking trail is $6\frac{2}{3}$ miles long. It has 4 exercise stations, spaced evenly along the trail. What is the distance between each exercise station?

   ____

4. Jamal buys a strip of 25 postage stamps. The strip of stamps is $21\frac{7}{8}$ inches long. How long is each stamp?

   ____

Choose the letter for the best answer.

5. Matt wants to decorate his skateboard with decals. His skateboard is $28\frac{3}{4}$ inches long. The decals are $5\frac{1}{2}$ inches long. If Matt arranges them in a line from the front to the back, how many decals will fit?

   A 5 decals
   B 6 decals
   C 7 decals
   D 8 decals

6. A square floor tile measures $\frac{3}{4}$ square feet. How many tiles are required to cover a 200 square foot floor?

   F 150 tiles
   G 267 tiles
   H 275 tiles
   J 300 tiles

7. Bev buys a sleeve of ball bearings for her skateboard. Each of the bearings is $1\frac{1}{5}$ inches wide. The sleeve is $9\frac{3}{5}$ inches long. How many bearings are in the sleeve?

   A 5 bearings
   B 8 bearings
   C 9 bearings
   D 12 bearings

8. The average hamster weighs about $\frac{1}{4}$ pound. The total weight of all the hamsters in a cage in a pet store is $1\frac{1}{2}$ pounds. How many hamsters are in the cage?

   F 3 hamsters
   G 4 hamsters
   H 5 hamsters
   J 6 hamsters

Name ______________________ Date ______________ Class ______________

LESSON 3-10

Reading Strategies

Use a Visual Model

The Smith family has a two-and-a-half-foot-long sandwich to share. One-half foot of the sandwich will serve one person. How many one-half foot servings are in this sandwich?

← $2\frac{1}{2}$ feet →

$\frac{1}{2}$	$\frac{1}{2}$	$\frac{1}{2}$	$\frac{1}{2}$	$\frac{1}{2}$

Use the model to answer each question.

1. How long is the sandwich?

____

2. How long is each serving?

____

3. If you divided the sandwich into $\frac{1}{2}$ ft servings, how many would you have?

____

4. What is $2\frac{1}{2} \div \frac{1}{2}$?

____

Suppose you have two sandwiches.

$\frac{1}{2}$	$\frac{1}{2}$	$\frac{1}{2}$	$\frac{1}{2}$	$\frac{1}{2}$

$\frac{1}{2}$	$\frac{1}{2}$	$\frac{1}{2}$	$\frac{1}{2}$	$\frac{1}{2}$

5. How many feet are in both sandwiches?

____

6. What is $2\frac{1}{2} \times 2$?

____

7. Compare the answers to $2\frac{1}{2} \div \frac{1}{2}$ and $2\frac{1}{2} \times 2$. What do you notice?

____

Name ______________________ Date ______________ Class ______________

LESSON 3-10

Puzzles, Twisters & Teasers

Stop Making Sense!

Decide whether or not each equation is correct. Circle the letter above your answer. Use the letters you circled to solve the riddle.

1. $\frac{2}{3} \div \frac{1}{5} = 3\frac{1}{3}$

   I correct B incorrect

2. $2 \div -\frac{7}{8} = 4\frac{1}{4}$

   D correct T incorrect

3. $\frac{3}{4} \div \frac{6}{7} = \frac{1}{3}$

   F correct M incorrect

4. $3\frac{3}{5} \div 9\frac{1}{7} = -3\frac{3}{5}$

   G correct A incorrect

5. $-4\frac{1}{3} \div 2\frac{1}{2} = -1\frac{11}{15}$

   K correct H incorrect

6. $\frac{3}{5} \div 6 = \frac{1}{10}$

   E correct J incorrect

7. $9 \div 1\frac{1}{2} = 6$

   S correct L incorrect

8. $10 \div \frac{5}{9} = \frac{3}{10}$

   O correct C incorrect

9. $-4\frac{4}{5} \div \frac{6}{7} = -5\frac{3}{5}$

   E correct P incorrect

10. $\frac{2}{3} \div \frac{8}{9} = 1$

    Q correct N incorrect

11. $\frac{3}{4} \div \frac{6}{7} = \frac{6}{8}$

    R correct T incorrect

12. $\frac{4}{9} \div 12 = 1\frac{3}{4}$

    V correct S incorrect

What did the little boy say when he learned how to count money?

___ ___

___ ___ ___ ___ ___

___ ___ ___ ___ ___!

Name ______________________ Date ______________ Class ______________

LESSON 3-11

Practice A

Solving Equations Containing Fractions

Solve. Choose the letter for the best answer.

1. $t - \frac{3}{4} = \frac{1}{4}$

A $t = \frac{1}{4}$ C $t = \frac{3}{4}$

B $t = \frac{1}{2}$ D $t = 1$

2. $g - \frac{3}{8} = \frac{1}{8}$

F $g = \frac{1}{4}$ H $g = \frac{3}{4}$

G $g = \frac{1}{2}$ J $g = 1$

3. $k + \frac{7}{12} = \frac{11}{12}$

A $k = \frac{1}{4}$ C $k = \frac{1}{2}$

B $k = \frac{1}{3}$ D $k = 1$

4. $n + \frac{2}{5} = \frac{4}{5}$

F $n = \frac{2}{5}$ H $n = \frac{4}{5}$

G $n = \frac{3}{5}$ J $n = 1$

5. $f + \frac{1}{6} = \frac{5}{6}$

A $f = \frac{1}{6}$ C $f = \frac{1}{2}$

B $f = \frac{1}{3}$ D $f = \frac{2}{3}$

6. $\frac{1}{4}s = 4$

F $s = \frac{1}{2}$ H $s = 4$

G $s = 1$ J $s = 16$

Solve. Write each answer in simplest form.

7. $p - \frac{1}{4} = \frac{1}{6}$

____

8. $d - \frac{2}{5} = \frac{3}{10}$

____

9. $y + \frac{5}{8} = \frac{3}{4}$

____

10. $\frac{3}{4}m = \frac{5}{6}$

____

11. $\frac{1}{2}x = \frac{5}{8}$

____

12. $\frac{5}{6}r = \frac{3}{10}$

____

13. Eunice walked $\frac{1}{4}$ mile from home to the store on her way to a friend's house. If the store is $\frac{1}{3}$ of the way to her friend's house, how far is her friend's house from home?

____

Name ______________________ Date ______________ Class ______________

LESSON 3-11

Practice B

Solving Equations Containing Fractions

Solve. Write each answer in simplest form.

1. $t - \frac{3}{7} = \frac{4}{7}$ ________

2. $g - \frac{5}{16} = \frac{3}{16}$ ________

3. $k - \frac{3}{10} = \frac{2}{5}$ ________

4. $n + \frac{1}{7} = \frac{2}{3}$ ________

5. $j + \frac{5}{6} = \frac{17}{18}$ ________

6. $f + \frac{5}{12} = \frac{3}{4}$ ________

7. $\frac{1}{4}s = \frac{3}{4}$ ________

8. $\frac{1}{5}a = \frac{1}{2}$ ________

9. $\frac{4}{5}h = \frac{8}{9}$ ________

10. $p - \frac{2}{3} = \frac{5}{8}$ ________

11. $d - \frac{3}{5} = \frac{7}{10}$ ________

12. $y - \frac{2}{7} = 3\frac{1}{4}$ ________

13. $c + \frac{5}{12} = 2\frac{1}{6}$ ________

14. $w + \frac{4}{15} = 3\frac{1}{3}$ ________

15. $z + \frac{6}{7} = 2\frac{3}{5}$ ________

16. $\frac{5}{6}m = \frac{8}{9}$ ________

17. $\frac{1}{2}x = 3\frac{7}{15}$ ________

18. $\frac{1}{5}r = 2\frac{2}{3}$ ________

19. Sarabeth ran $1\frac{2}{5}$ miles on a path around the park. This was $\frac{5}{8}$ of the distance around the park. What is the distance around the park?

____

20. An interior decorator bought $12\frac{1}{2}$ yards of material to make drapes. He used $8\frac{2}{3}$ yards on 1 pair of drapes. How much material does he have left?

____

Name ____________________ Date ______________ Class ______________

LESSON 3-11

Practice C

Solving Equations Containing Fractions

Solve. Write each answer in simplest form.

1. $t - \frac{12}{19} = \frac{5}{19}$ ________

2. $g - \frac{5}{22} = \frac{3}{11}$ ________

3. $k - \frac{3}{4} = \frac{2}{3}$ ________

4. $n + \frac{8}{15} = \frac{3}{10}$ ________

5. $j + \frac{11}{12} = \frac{13}{18}$ ________

6. $f + \frac{8}{11} = \frac{4}{5}$ ________

7. $\frac{1}{7}s = 1\frac{1}{4}$ ________

8. $\frac{3}{5}a = 3\frac{2}{7}$ ________

9. $-\frac{4}{9}h = -6\frac{2}{7}$ ________

10. $p - \frac{2}{7} = 2\frac{3}{8}$ ________

11. $d + \frac{5}{12} = -2\frac{2}{15}$ ________

12. $y - \frac{2}{5} = 4\frac{4}{9}$ ________

13. $-\frac{7}{10} + c = 5\frac{3}{20}$ ________

14. $w + 1\frac{7}{11} = 4\frac{2}{5}$ ________

15. $z + 3\frac{1}{6} = 6\frac{2}{9}$ ________

16. $\frac{5}{8}m = 2\frac{4}{7}$ ________

17. $-\frac{1}{11}x = 2\frac{1}{10}$ ________

18. $\frac{3}{7}r = 3\frac{1}{3}$ ________

19. A grizzly bear can run $1\frac{1}{5}$ times as fast as an elephant. A grizzly bear can run 30 mi/h. How fast can an elephant run?

____

20. A fruit salad recipe with bananas, strawberries, and grapes requires $3\frac{1}{2}$ pounds of fruit. You need $1\frac{3}{4}$ pounds of bananas and $\frac{7}{8}$ of a pound of strawberries. How many pounds of grapes are needed?

____

Name ______________________ Date ______________ Class ______________

LESSON 3-11

Review for Mastery

Solving Equations Containing Fractions

You can use addition to solve a subtraction equation involving fractions.

$$x - \frac{4}{9} = \frac{1}{3}$$

$$x - \frac{4}{9} + \mathbf{\frac{4}{9}} = \frac{1}{3} + \mathbf{\frac{4}{9}}$$

$$x = \frac{3}{9} + \frac{4}{9}$$

$$x = \frac{7}{9}$$

Remember, addition undoes subtraction.

You can use subtraction to solve an addition equation involving fractions.

$$n + \frac{2}{5} = \frac{9}{10}$$

$$n + \frac{2}{5} - \mathbf{\frac{2}{5}} = \frac{9}{10} - \mathbf{\frac{2}{5}}$$

$$n = \frac{9}{10} - \frac{4}{10}$$

$$n = \frac{5}{10} = \frac{1}{2}$$

Remember, subtraction undoes addition.

Solve. Write each answer in simplest form.

1. $d - \frac{1}{6} = \frac{3}{4}$

$$d - \frac{1}{6} + \frac{}{\_\_} = \frac{3}{4} + \frac{}{\_\_}$$

$$d = \frac{\_\_}{12} + \frac{\_\_}{12}$$

$$d = \frac{\_\_}{12}$$

2. $y + \frac{4}{5} = \frac{14}{15}$

$$y + \frac{4}{5} - \frac{}{\_\_} = \frac{14}{15} - \frac{}{\_\_}$$

$$y = \frac{14}{15} - \frac{\_\_}{15}$$

$$y = \frac{\_\_}{15}$$

3. $t - \frac{1}{8} = \frac{3}{4}$

____

4. $k + \frac{1}{2} = 1\frac{5}{8}$

____

5. $a - \frac{3}{5} = \frac{7}{10}$

____

LESSON 3-11

Review for Mastery

Solving Equations Containing Fractions (continued)

You can use division to solve a multiplication equation involving fractions. Multiply both sides of the equation by the reciprocal of the coefficient of the variable.

$$3y = \frac{9}{10}$$

$$3y \cdot \frac{1}{3} = \frac{9}{10} \cdot \frac{1}{3}$$

The reciprocal of 3 is $\frac{1}{3}$.

$$y = \frac{9}{10} \cdot \frac{1}{3}$$

$$y = \frac{9}{30} = \frac{3}{10}$$

$$\frac{5}{6}a = \frac{1}{2}$$

$$\frac{5}{6}a \cdot \frac{6}{5} = \frac{1}{2} \cdot \frac{6}{5}$$

The reciprocal of $\frac{5}{6}$ is $\frac{6}{5}$.

$$a = \frac{1}{2} \cdot \frac{6}{5}$$

$$a = \frac{6}{10} = \frac{3}{5}$$

Solve. Write each answer in simplest form.

6. $8x = 3\frac{1}{5}$

$$8x = \frac{\_\_}{5}$$

$$8x \cdot \frac{\_\_}{\_\_} = \frac{\_\_}{5} \cdot \frac{\_\_}{\_\_}$$

$$x = \frac{\_\_}{5} \cdot \frac{\_\_}{\_\_}$$

$$x = \frac{\_\_}{\_\_} = \frac{\_\_}{\_\_}$$

7. $\frac{2}{3}k = \frac{5}{6}$

$$\frac{2}{3}k \cdot \frac{\_\_}{\_\_} = \frac{5}{6} \cdot \frac{\_\_}{\_\_}$$

$$k = \frac{5}{6} \cdot \frac{\_\_}{\_\_}$$

$$k = \frac{\_\_}{\_\_} = \frac{\_\_}{\_\_} = \_\_$$

8. $\frac{3}{4}d = 5$ ____________

9. $6y = \frac{2}{3}$ ____________

10. $\frac{1}{5}s = \frac{5}{8}$ ____________

Name ______________________ Date ______________ Class ______________

LESSON 3-11

Challenge

Shopping List Equation

On the right is the list of ingredients for a recipe for bean and cheese tacos. The recipe serves 4 people. You are making bean and cheese tacos for 18 people.

Use the ingredients to answer the questions and solve the equations.

Bean and Cheese Tacos

$\frac{1}{2}$ pound kidney beans

1 clove garlic, chopped

4 flour tortillas

1 cup ricotta cheese

$\frac{1}{4}$ cup parmesan cheese

$\frac{1}{4}$ cup green onions, chopped

Serves 4

1. By what factor do you need to multiply each ingredient to serve 18 people? Solve the equation to find out. $4x = 18$

   $x =$ ______________________

2. How many pounds of kidney beans do you need to serve 18 people?

   ____

3. Kidney beans come in $\frac{3}{4}$-pound cans. How many cans do you need to serve 18 people? Write and solve an equation.

   ____

4. Ricotta cheese is on sale in $1\frac{1}{4}$-cup tubs. How many tubs do you need to serve 18 people? Write and solve an equation.

   ____

5. You need $1\frac{1}{8}$ cups of Parmesan cheese. You have $\frac{1}{4}$ cup. Write and solve an equation to find how much more you need.

   ____

6. There are 4 green onions in $\frac{1}{4}$ cup. Write and solve an equation to find how many green onions you need.

   ____

7. You have 8 cloves of garlic and you need $4\frac{1}{2}$. How many cloves of garlic will you have left? Write and solve an equation.

   ____

Name ______________________ Date ______________ Class ______________

LESSON 3-11

Problem Solving

Solving Equations Containing Fractions

Write the correct answer.

1. At the 2002 Winter Olympics, Austria won 2 gold medals. This was $\frac{1}{8}$ of the total medals Austria won. How many medals did Austria win?

____

2. At the 2002 Winter Olympics, Germany won 35 medals, of which 16 were silver. They won $1\frac{1}{3}$ times as many silver medals as gold medals. How many gold medals did Germany win?

____

3. Jesse plays ice hockey. He is on the ice $\frac{2}{5}$ of each game. A game lasts 45 minutes, divided into 3 periods. How many minutes is Jesse on the ice during each game?

____

4. Amelia's soccer team won $\frac{3}{4}$ of its games. The team won 18 games. How many games did the team play?

____

Choose the letter for the best answer.

5. At the 2002 Winter Olympics, China won 8 medals. The United States won $4\frac{1}{4}$ times as many medals as China did. How many medals did the United States win?

   A 24 medals
   B 32 medals
   C 34 medals
   D 40 medals

6. Juanita's water bottle contains $1\frac{7}{8}$ liters. She uses her bottle to fill Felix's. When she is done, her bottle contains $\frac{1}{4}$ liter. How many liters can Felix's bottle hold?

   F $\frac{7}{8}$ liter
   G $1\frac{1}{8}$ liters
   H $1\frac{5}{8}$ liters
   J $2\frac{1}{8}$ liters

7. Yoriko ran $6\frac{1}{2}$ laps of the track. Each lap is $\frac{1}{4}$ of a mile. How many miles did she run?

   A $6\frac{5}{8}$ miles
   B $4\frac{1}{2}$ miles
   C 2 miles
   D $1\frac{5}{8}$ miles

8. Rocky runs $3\frac{1}{2}$ miles each week. Leroy runs $5\frac{1}{3}$ miles each week. How much farther does Leroy run?

   F $1\frac{1}{2}$ miles
   G $1\frac{5}{6}$ miles
   H $2\frac{1}{6}$ miles
   J $2\frac{1}{4}$ miles

Name ______________________ Date ______________ Class ______________

LESSON 3-11

Reading Strategies

Compare and Contrast

Compare the steps for solving equations with fractions and solving equations with whole numbers.

Steps for Solving Equations		
	Whole Numbers	**Fractions**
Step 1: Get *x* by itself on one side of the equation.	$x - 8 = 7$	$x - \frac{3}{12} = \frac{4}{12}$
Step 2: Perform the opposite operation. In a subtraction problem, you add to get *x* by itself.	$x - 8 + \mathbf{8} = 7 + \mathbf{8}$	$x - \frac{3}{12} + \frac{\mathbf{3}}{\mathbf{12}} = \frac{4}{12} + \frac{\mathbf{3}}{\mathbf{12}}$
Step 3: Solve.	$x = 15$	$x = \frac{7}{12}$

Use the chart to answer each question.

1. What is the first step to solve an equation with whole numbers?

____

2. Compare the first step in solving an equation with whole numbers to fractions. Is it the same or different?

____

3. What is the second step in solving an equation with whole numbers?

____

4. Compare the second step in solving an equation with whole numbers to an equation with fractions. Is it the same or different?

____

5. What is the opposite operation in both of the examples in the chart?

____

6. What is the third step in solving an equation with whole numbers?

____

7. Compare the third step in solving an equation with whole numbers to solving an equation with fractions. Is it the same or different?

____

Name ______________________ Date ____________ Class ____________

LESSON 3-11

Puzzles, Twisters & Teasers

What's Up, Doc?

Solve the equations. Then use the letters of the variables to answer the riddle.

1. $\frac{7}{18} + t = -\frac{4}{7}$ **t** = ____________

2. $e - \frac{1}{5} = \frac{3}{5}$ **e** = ____________

3. $g + \frac{11}{20} = \frac{3}{4}$ **g** = ____________

4. $\frac{2}{3}k = \frac{4}{5}$ **k** = ____________

5. $3a = \frac{6}{7}$ **a** = ____________

6. $\frac{5}{12} + o = \frac{2}{3}$ **o** = ____________

7. $m - \frac{1}{2} = \frac{1}{4}$ **m** = ____________

8. $\frac{1}{5}r = 8$ **r** = ____________

9. $s - \frac{2}{3} = \frac{5}{6}$ **s** = ____________

10. $\frac{2}{3}x = \frac{3}{5}$ **x** = ____________

Why did the rabbit buy a pure gold ring?

$-\frac{121}{126}$ $\frac{1}{4}$ $\frac{1}{5}$ $\frac{4}{5}$ $-\frac{121}{126}$

$\frac{3}{4}$ $\frac{1}{4}$ 40 $\frac{4}{5}$

$1\frac{1}{5}$ $\frac{2}{7}$ 40 $\frac{2}{7}$ $-\frac{121}{126}$ $1\frac{1}{2}$

LESSON 3-1

Practice A

1. 4 2. 2
3. 11 4. 6
5. 12 6. 19

Estimates may vary.

7. 24 8. 8
9. 17 10. 3
11. –150 12. 1
13. 13 14. 15
15. 19

Possible answers given.

16. 8 17. 12
18. 28 19. 55
20. 150 21. 45
22. 4 23. 11
24. 18 25. about 6 gallons

Practice B

Estimates may vary.

1. 43 2. 41
3. 21 4. 15
5. 6 6. 2
7. –53 8. 31
9. 66 10. –9
11. 43 12. –75

Possible answers given.

13. 3 14. 860
15. 1,100 16. –160
17. 210 18. 3
19. 72 20. 11
21. 15 22. $20
23. about 6 months

Practice C

Estimates may vary. Possible answers given.

1. 68 2. 58
3. 4 4. 0
5. 840 6. 144
7. –921 8. –490
9. –30 10. 1,710
11. –52 12. 2,700
13. 261 14. –17
15. 18 16. –645
17. 20 18. –126
19. –8 20. 600
21. 33 22. 5
23. –8 24. –720
25. 195 miles 26. no; about $60

Review for Mastery

1. ones
8
tens
50
8 + 50 = 58

2. tens
30
tens
10
30 – 10 = 20

3. ones
6
tens
–20
6 (–20) = –120

4. tens
40
ones
9
40 ÷ 9; 45 ÷ 9
45 ÷ 9 = 5

5. 120 6. 6
7. 50 8. 150
9. –280 10. 5

Challenge

1. > 2. >
3. < 4. <
5. < 6. >
7. > 8. <
9. < 10. <
11. < 12. <
13. < 14. <
15. < 16. >

17. > 18. <

19. > 20. <

21. > 22. <

23. < 24 >

Problem Solving

Estimates may vary. Possible answers given.

1. more than 20 pies
2. yes; about $3.00
3. no; about 25 MB
4. 2 CDs 5. D
6. G 7. H

Reading Strategies

1. round down, because 4 is less than 5
2. round up, because 6 is greater than 5
3. 7 + 3 = 10 4. 14
5. 5
6. It is easier to solve 14 − 5 because there are fewer digits to subtract.

Puzzles, Twisters & Teasers

I. 2

D. 250

L. 63

N. −23

G. 92

U. 7

O. 8

R. 5

A. −7

H. 143

E. 28

Y. −60

I' L L G O O N

A H E A D, Y O U

H A N G

A R O U N D

LESSON 3-2

Practice A

1. 9; 8; 5 2. 2; 2; 2; 3
3. 1; 1; 5; 7 4. 2; 6; 5
5. 2.36 6. 31.299
7. 33.176 8. 6; 6; 7
9. 2; 0; 6 10. 3.8
11. 5.4 12. 5.2
13. 7.12 14. 3.5
15. 9.68

Practice B

1. 15.05 2. 17.59
3. 33.62 4. 15.72
5. −16 6. −24.73
7. 14.68 8. −6.96
9. 90.9 10. 2.61
11. 1.8 12. 19.46
13. 16.73 14. 9.05
15. 19.71 16. 12.08
17. 10.11 18. 1.55
19. 45.93 20. 3.37
21. 29.32 22. 80.83 seconds
23. 0.83 ounce

Practice C

1. 16.816 2. 2.975
3. 46.718 4. 1.175
5. 12.979 6. −29.491
7. −2.716 8. 23.392
9. −81.632 10. 38.026
11. −30.774 12. −7.259
13. 24.591 14. 111.105
15. −81.054 16. 17.99
17. 36.054 18. 36.794
19. 54.732 20. 90.559
21. 66.837 22. 39.581
23. 39.35 mi/h
24. 24.355 kilograms

Review for Mastery

1. 28.12
2. 2.07
3. 15.66
4. 49.96
5. 43.9
6. 29.68
7. 37.57

Challenge

1. −61.244 + 59.014 −2.23	2. 21.65 + 14.2 + 10.03 45.88	3. 41.729 + 14.92 59.649
4. 4.018 + 7.052 11.07	5. 212.8 − 136.2 76.6	6. 42.25 − 13.174 29.076
7. −104.02 + 149.9 45.88	8. 0.015 − 6.72 −6.705	9. −0.673 + (25.707) −6.38
10. −5.23 − 0.804 + 35.11 29.076	11. −70.6 + 48.05 + 33.62 11.07	12. 22.015 + 23.865 45.88
13. 60.25 − 0.601 59.649	14. 65.109 − 5.46 59.649	15. 7.17 − 9.4 −2.23
16. −1.08 + 3.797 − 9.422 −6.705	17. 10.25 − 12.03 − 0.45 −2.23	18. 16.02 + 44.48 + 16.1 76.6
19. 15.25 + 4.03 + 17.706 36.986	20. 130.82 − 101.744 29.076	21. −5.61 − 1.095 −6.705
22. 11.691 − 21.1 + 3.029 −6.38	23. 45.006 + (−8.02) 36.986	24. 52.73 − 53.71 + 12.05 11.07
25. −33.42 − (−110.02) 76.5	26. 46.01 + 9.9 − 62.29 −6.38	27. 6.007 + 74.2 − 43.221 36.986

Problem Solving

1. 3.93 kilometers
2. 4.2 kilometers
3. 4.39 kilometers
4. −4.9 °C
5. A
6. H
7. B

Reading Strategies

1. One number is placed in each square.
2. as a placeholder to show that there is no number in that place
3.

4	0	.	3	0
−	6	.	5	4

4. yes; in the hundredths place of the first number
5. 33.76

Puzzles, Twisters & Teasers

R. 22
I. 3
E. 13
L. 10
N. 2
M. 39
O. 15
Y. 19
C. 8
U. 4
H. 0
A. −5
T. 30

I'LL MEET YOU AT THE CORNER

LESSON 3-3

Practice A

1. C
2. G
3. D
4. G
5. A
6. H
7. A
8. H
9. 27.2
10. −16.2
11. 47.5
12. 24.5
13. −37.8
14. 33.3
15. 33.5
16. −70.4
17. −20.8
18. −12.3
19. 1.8
20. −3.91
21. 29.6 miles

Practice B

1. 1.8
2. 0.15
3. 2.8
4. 48.8
5. 44.4
6. 12.6
7. 33.6
8. 24.6
9. 44
10. 9
11. 15.8
12. 34.5
13. 49.14
14. −45.59

15. 6.12
16. 22.96
17. –44.94
18. 13.718
19. 44.52
20. 0.2148
21. –12.168
22. 14.26
23. 0.899
24. –10.906
25. 100.8 miles

Practice C

1. 34.1404	2. 32.6876
3. 40.0475	4. 8.8922
5. 72.3632	6. 29.7032
7. 359.2032	8. 92.1088
9. 767.0376	10. 20.538
11. –30.5778	12. 60.0625
13. 49.8225	14. –64.5696
15. –26.8732	16. 377.2725
17. –735.7801	18. –99.8138
19. –28.329	20. 0.0486
21. –213.8412	22. –1.0727
23. –943.2126	24. 32.472
25. $29.13	26. $42.35

Review for Mastery

1. 7
   9
   63
   60.97
2. –3
   9
   –27
   –28.248
3. 4
   1
   4
   3.28
4. 12
   –3
   –36
   –33.21
5. 16.274
6. –2.934
7. –34.566

Challenge

1. 216;
   17.5 • 11.5 = 201.25;
   18.4 • 12.4 = 228.16
2. 13.2;
   3.25 • 3.5 = 11.375;
   3.34 • 4.4 = 14.696
3. 0.63;
   0.85 • 0.65 = 0.5525;
   0.94 • 0.74 = 0.6956
4. 3.96;
   1.75 • 2.15 = 3.7625;
   1.84 • 2.24 = 4.1216
5. 0.6;
   5.5 • 0.05 = 0.275;
   6.4 • 0.14 = 0.896
6. 2.7;
   4.45 • 0.55 = 2.4475;
   4.54 • 0.64 = 2.9056
7. 11;
   9.5 • 1.05 = 9.975;
   10.4 • 1.14 = 11.856
8. 4.5;
   0.25 • 14.5 = 3.625;
   0.34 • 15.4 = 5.236
9. 44.37;
   8.65 • 5.05 = 43.6825;
   8.74 • 5.14 = 44.9236
10. 2.66;
   1.35 • 1.85 = 2.4975;
   1.44 • 1.94 = 2.7936

Problem Solving

1. $105
2. $83.25
3. $6.00
4. about 88.45 million tickets
5. C
6. H
7. D
8. G

Reading Strategies

1. The same numerals are in both answers.
2. There are decimal places in the product when you multiply decimals, but none with whole numbers.
3. 2
4. 3

5. 5

Puzzles, Twisters & Teasers

S. 3
O. 5
U. 29
T. 8
I. 10
R. 0
N. 21
A. –1
G. 6
H. 20
E. –4
W. 2

WHEN IT TURNS INTO A GARAGE

LESSON 3-4

Practice A

1. 2
2. 60
3. 30
4. –3
5. 3.5
6. –4.1
7. 5
8. –3.75
9. 15
10. 4
11. 2.5
12. –52
13. 27 miles per gallon
14. 30 hours

Practice B

1. 24
2. 6.2
3. –440
4. 66.5
5. –6.6
6. –161.5
7. –14
8. 2.5
9. 12
10. –22.5
11. 8.75
12. 250
13. 23 miles per gallon
14. 9 members

Practice C

1. –7.84
2. 74
3. –75
4. 5.42
5. –8.91
6. 37.5
7. 627.3
8. –495.8
9. –573
10. 20
11. –2.9
12. –4.5
13. 6
14. 30.72
15. 3.9
16. 12 years

Review for Mastery

1. 0.8
   a. 1
   b. 1
   c. $56\overline{)44.8}$
   d. 0.8
2. 0.7
3. 405
4. 1.2
5. 20
   a. 2
   b. 2
   c. 2
   d. 20
6. 5
7. 160
8. 38.4

Challenge

1. Check students' drawings; rectangle should be about 5.3 cm long and 3.3 cm wide.
2. 1.62
3. 4.94 meters
4. 2.99 inches
5. 3.76 feet
6. 7.41 yards
7. 10.19 centimeters
8. 15.19 meters
9. 33.95 inches
10. 51.36 feet
11. 60.62 yards
12. 71.60 centimeters
13. 136.11 inches
14. 215.25 meters

Problem Solving

1. 5 pieces
2. 0.162 kilogram
3. 79 seconds
4. 12 squares
5. D
6. H
7. A
8. H

Reading Strategies

1. Possible answer: The numerals are the same.
2. Possible answer: One equation uses decimal values. The other uses whole numbers.
3. They are both 5.
4. Move the decimal point one place to the right.
5. Move the decimal point one place to the right.
6. no

Puzzles, Twisters & Teasers

D; O; Y

O; U; H

E; A; R

A; P; H

O; N; E

DO YOU HEAR A PHONE

Isabel Ringing

LESSON 3-5

Practice A

1. B
2. H
3. B
4. G
5. $x = 9.9$
6. $h = 5.8$
7. $k = -16.8$
8. $g = 6.84$
9. $w = 0.57$
10. $m = 12.75$
11. $n = 3.5$
12. $j = 6$
13. $p = -0.21$
14. $y = 16$
15. $x = 9.5$
16. $m = 3.2$
17. $s = 32.4$
18. $f = 5.6$
19. $d = 3.22$
20. $c = 0.7$
21. $h = 73.2$
22. $a = 2.94$
23. $6.25
24. $47.20

Practice B

1. $t = 8.53$
2. $p = -10.46$
3. $r = 5.66$
4. $e = -5.05$
5. $w = 24.97$
6. $z = 24.33$
7. $p = 92.31$
8. $h = -14.16$
9. $k = -136.82$
10. $x = 6.42$
11. $n = 14.64$
12. $y = 7.05$
13. $p = 6.7$
14. $j = 9.9$
15. $g = -27.4$
16. $y = 6.8$
17. $x = 14.7$
18. $c = 16.7$
19. $e = 51.06$
20. $f = 198.12$
21. $d = 201.76$
22. $w = -3.08$
23. $m = 16.66$
24. $s = -74.62$
25. $19.25

Practice C

1. $t = 4.61$
2. $p = -15.74$
3. $c = 7.8$
4. $e = -14.39$
5. $w = 7.9$
6. $g = 7.1$
7. $a = 32.94$
8. $h = -15.25$
9. $k = -137.34$
10. $r = 25.7$
11. $j = 5.7$
12. $p = -8.4$
13. $y = 3.9$
14. $x = 8.8$
15. $z = 13.39$
16. $a = 50.246$
17. $b = -23.581$
18. $d = 201.566$
19. $s = 2.54$
20. $v = 5.9274$
21. $f = 193.802$
22. $n = 2.2382$
23. $t = 11.656$
24. $w = -0.02632$
25. 3.8 miles
25. –$7,429

Review for Mastery

1. 2.2
2. 1.9; 1.9
7.3
3. +8.3; +8.3
4.18
4. –5.75; –5.75
–11.95
5. $t = 14.6$
6. $p = 33.1$
7. $r = -11.05$
8. $k = -21.7$

9. 5.7; 5.7
   8

11. 0.14; 0.14
    2.1

13. $y = 11.7$

15. $s = 39.9$

10. ÷; −6; ÷; −6 ;
    −1.34

12. 6.3; •; •; 6.3
    −57.33

14. $c = -6.1$

16. $h = -2.88$

Challenge

1. (12.7, −15.4)
2. (2, 19.2)
3. (3.12, 0.78)
4. (0.85, 4.25)
5. (0.4, 0.2)
6. (1.98, 6.93)
7. (−1.2, 2.2)
8. (−2.7, 3.3)
9. (−4.3, −2.15)
10. (−10.4, 21.9)
11. (1.4, −4.2)
12. (−1.28, 1.63)

Problem Solving

1. 2.4 meters
2. 14.793 grams
3. 5.1 miles
4. 47.5 pounds
5. B
6. G
7. D
8. J

Reading Strategies

1. The steps are the same.
2. Whole numbers are used in whole number equations and decimals are used in decimal equations.
3. Get *y* by itself on one side of the equation.
4. division
5. Divided by 3 in both types of equations.

Puzzles, Twisters & Teasers

g. −2
t. 0.9
n. −0.15
h. 0.72
i. −4.8
e. −7.11
l. 3.92
y. −0.9
b. −1.2
a. 6.48
r. 16
w. 2.5
d. −5.44
s. 21.61
x. −53.21

THEY'RE ALWAYS DRIBBLING

LESSON 3-6

Practice A

Possible answers given.

1. $\frac{1}{2}$; $\frac{1}{2}$
2. 5; 2; 3
3. $1\frac{1}{2}$
4. $\frac{1}{2}$
5. 6
6. $\frac{1}{2}$
7. 0
8. $4\frac{1}{2}$
9. 10; 80
10. 40; 8; 5
11. 24
12. 20
13. 54
14. 4
15. 3
16. 3
17. 9 miles
18. $11\frac{1}{2}$ miles

Practice B

Possible answers given.

1. 1
2. 2
3. $\frac{1}{2}$
4. 1
5. $3\frac{1}{2}$
6. 1
7. $7\frac{1}{2}$
8. $5\frac{1}{2}$
9. 6
10. 2
11. 40
12. 15
13. 6
14. 2

15. 63
16. 72
17. 9
18. 34
19. 64 square feet
20. 9 pounds

Practice C

Possible answers given.

1. $1\frac{1}{2}$
2. 0
3. $1\frac{1}{2}$
4. 6
5. 4
6. $4\frac{1}{2}$
7. $6\frac{1}{2}$
8. $8\frac{1}{2}$
9. $9\frac{1}{2}$
10. 36
11. 3
12. –4
13. –28
14. 51
15. 3
16. 48
17. 16
18. –4
19. 107 pounds
20. 128 cubic feet

Review for Mastery

1. 1

   $\frac{1}{2}$

   1; $\frac{1}{2}$

   $\frac{1}{2}$

2. 4

   2

   4; 2

   2

3. $1\frac{1}{2}$
4. $1\frac{1}{2}$
5. 10
6. 18
7. 3
8. 30

Challenge

1. $1\frac{6}{11}$; $-\frac{9}{10}$
2. $-4\frac{5}{6}$
3. $\frac{6}{11}$; $\frac{4}{19}$; $-4\frac{16}{19}$

H I G H E R

4. $\frac{9}{10}$; $-6\frac{9}{10}$
5. $-6\frac{5}{9}$
6. $2\frac{1}{3}$

W A V Y

7. $6\frac{1}{8}$; $8\frac{3}{10}$
8. $29\frac{7}{8}$; $4\frac{11}{13}$
9. $8\frac{4}{5}$

S T A M P

Problem Solving

1. about 4 feet
2. about $2\frac{1}{2}$ jugs
3. about 9 miles
4. about 36 inches
5. B
6. H
7. B
8. F

Reading Strategies

1. close to $\frac{1}{2}$
2. close to 0
3. close to 1
4. close to 1
5. close to 0
6. close to 0
7. close to $\frac{1}{2}$

Puzzles, Twisters & Teasers

H. 7

T. 5

S. 8

E. 6

I. $\frac{1}{2}$

Y. $1\frac{1}{2}$

F. 25
A. 3
L. 26
R. 1

THEY ARE SHELLFISH

LESSON 3-7

Practice A

1. 1
2. $\frac{4}{5}$
3. $\frac{1}{2}$
4. $\frac{3}{4}$
5. $1\frac{2}{3}$
6. $1\frac{1}{5}$
7. $\frac{1}{2}$
8. $\frac{1}{4}$
9. $-\frac{3}{5}$
10. $\frac{1}{4}$
11. $-\frac{1}{3}$
12. $\frac{2}{5}$
13. $1\frac{1}{12}$
14. $1\frac{1}{12}$
15. $\frac{19}{24}$
16. $-\frac{7}{12}$
17. $\frac{11}{20}$
18. $\frac{3}{10}$
19. $\frac{1}{4}$ pound

Practice B

1. $\frac{3}{5}$
2. $\frac{4}{5}$
3. $\frac{1}{6}$
4. $\frac{1}{5}$
5. $-\frac{1}{3}$
6. $1\frac{1}{7}$
7. $1\frac{1}{5}$
8. $\frac{1}{8}$
9. $\frac{13}{21}$
10. $\frac{1}{20}$
11. $\frac{7}{8}$
12. $-\frac{11}{25}$
13. $1\frac{3}{4}$
14. $\frac{11}{24}$
15. $1\frac{11}{15}$
16. $-\frac{19}{40}$
17. $1\frac{17}{30}$
18. $\frac{13}{60}$
19. $\frac{5}{24}$ mile
20. $\frac{3}{8}$

Practice C

1. $\frac{1}{5}$
2. 1
3. $1\frac{5}{21}$
4. $\frac{13}{15}$
5. $-\frac{1}{36}$
6. $-\frac{2}{3}$
7. $\frac{11}{18}$
8. $1\frac{2}{25}$
9. $\frac{73}{88}$
10. $-\frac{7}{20}$
11. $1\frac{5}{24}$
12. $\frac{5}{18}$
13. $2\frac{1}{8}$
14. $-\frac{7}{12}$
15. $-\frac{3}{5}$
16. $-\frac{1}{30}$
17. $\frac{41}{45}$
18. $\frac{19}{28}$
19. $\frac{1}{21}$ billion
20. $1\frac{1}{30}$ miles
21. about $\frac{43}{50}$

Review for Mastery

1. $\frac{9}{15}$; $\frac{5}{15}$; $\frac{9+5}{15}$; $\frac{14}{15}$
2. $\frac{3}{9}$; $\frac{8-3}{9}$; $\frac{5}{9}$

3. $\frac{4}{10}$; $\frac{5}{10}$; $\frac{4+5}{10}$; $\frac{9}{10}$

4. $\frac{9}{12}$; $\frac{4}{12}$; $\frac{9-4}{12}$; $\frac{5}{12}$

5. $\frac{8}{12}$; $\frac{9}{12}$; $\frac{8+9}{12}$; $\frac{17}{12}$; $1\frac{5}{12}$

6. $\frac{2}{8}$; $\frac{5}{8}$; $\frac{2-5}{8}$; $-\frac{3}{8}$

7. $\frac{11}{12}$

8. $\frac{2}{5}$

9. $1\frac{1}{10}$

10. $\frac{1}{4}$

11. $-\frac{9}{20}$

12. $1\frac{11}{20}$

Challenge

1. $1\frac{5}{8}$

2. $-\frac{5}{7}$

3. $\frac{23}{36}$

$\frac{1}{2}$	$\frac{1}{7}$	$\frac{23}{36}$
$\frac{2}{12}$	$-\frac{5}{7}$	$\frac{5}{7}$
$1\frac{5}{8}$	$-\frac{1}{6}$	$\frac{7}{8}$

4. $-\frac{1}{12}$

5. $-1\frac{5}{18}$

6. $\frac{3}{5}$

$\frac{1}{12}$	1	$-\frac{3}{5}$
$\frac{5}{12}$	$1\frac{5}{18}$	$\frac{4}{5}$
$-\frac{1}{12}$	$-1\frac{5}{18}$	$\frac{3}{5}$

7. $1\frac{3}{7}$

8. $-\frac{1}{15}$

9. $-\frac{14}{45}$

$-\frac{1}{7}$	$1\frac{3}{7}$	1
$\frac{1}{15}$	$-\frac{1}{15}$	$\frac{1}{7}$
$-\frac{2}{45}$	$-\frac{14}{45}$	$\frac{2}{45}$

Problem Solving

1. $\frac{11}{20}$ of the total

2. $\frac{1}{2}$ of the budget

3. $\frac{5}{24}$ yard

4. Levi

5. C

6. G

7. B

8. F

Reading Strategies

1. $\frac{3}{8}$

2. $\frac{2}{8}$

3. $\frac{5}{8}$

4. no

5. $\frac{4}{6}$

6. $\frac{3}{6}$

7. $\frac{1}{6}$

8. numerator

Puzzles, Twisters & Teasers

D. $-\frac{1}{4}$

B. $\frac{13}{18}$

E. $-\frac{1}{12}$

C. $1\frac{23}{35}$

O. $1\frac{1}{21}$

A. $1\frac{3}{20}$

L. $\frac{3}{14}$ U. 1

S. $1\frac{1}{12}$ T. $\frac{3}{8}$

I. $\frac{13}{15}$ W. $-\frac{1}{6}$

H. $\frac{2}{3}$ N. $1\frac{1}{2}$

BECAUSE THE WINDOW IS CLOSED

LESSON 3-8

Practice A

1. $7\frac{3}{4}$
2. 8
3. $3\frac{3}{5}$
4. $4\frac{7}{8}$
5. $8\frac{3}{7}$
6. $7\frac{1}{9}$
7. $5\frac{13}{24}$
8. $8\frac{7}{12}$
9. $3\frac{7}{24}$
10. $2\frac{1}{2}$
11. $2\frac{2}{3}$
12. $3\frac{1}{3}$
13. $3\frac{1}{2}$
14. $\frac{2}{3}$
15. $2\frac{1}{3}$
16. $3\frac{7}{12}$
17. $2\frac{11}{12}$
18. $3\frac{5}{8}$
19. $3\frac{5}{8}$ miles

Practice B

1. 14
2. $9\frac{2}{9}$
3. $12\frac{7}{12}$
4. $6\frac{1}{5}$
5. $8\frac{3}{10}$
6. $4\frac{3}{20}$
7. $8\frac{21}{40}$
8. $6\frac{19}{24}$
9. $7\frac{4}{9}$
10. $4\frac{2}{9}$
11. $4\frac{2}{5}$
12. $3\frac{2}{5}$
13. $3\frac{1}{15}$
14. $4\frac{5}{12}$
15. $\frac{17}{40}$
16. $2\frac{1}{4}$
17. $1\frac{39}{40}$
18. $1\frac{3}{28}$
19. $9\frac{1}{8}$ miles

Practice C

1. $5\frac{19}{21}$
2. $8\frac{1}{26}$
3. $5\frac{5}{11}$
4. $7\frac{9}{19}$
5. $2\frac{3}{25}$
6. $1\frac{1}{4}$
7. $15\frac{29}{40}$
8. $11\frac{2}{21}$
9. $12\frac{19}{24}$
10. $3\frac{7}{30}$
11. $2\frac{17}{20}$
12. $4\frac{17}{18}$
13. $1\frac{1}{30}$
14. $13\frac{7}{16}$
15. $-20\frac{1}{12}$
16. $5\frac{19}{20}$
17. $2\frac{37}{40}$
18. $-11\frac{17}{35}$

19. < 20. <

21. = 22. >

23. $4\frac{3}{4}$ feet 24. $2\frac{1}{2}$ miles

Review for Mastery

1. $\frac{47}{10}$; $\frac{29}{10}$; $\frac{76}{10}$; $7\frac{6}{10}$; $7\frac{3}{5}$

2. $\frac{5}{2}$; $\frac{11}{8}$; $\frac{20}{8}$; $\frac{11}{8}$; $\frac{31}{8}$; $3\frac{7}{8}$

3. $\frac{16}{5}$; $\frac{7}{3}$; $\frac{48}{15}$; $\frac{35}{15}$; $\frac{83}{15}$; $5\frac{8}{15}$

4. $6\frac{5}{7}$ 5. $9\frac{1}{4}$

6. $7\frac{1}{9}$ 7. $6\frac{3}{10}$

8. $4\frac{7}{12}$ 9. $6\frac{5}{6}$

10. $\frac{25}{4}$; $\frac{7}{4}$; $\frac{18}{4}$; $4\frac{2}{4}$; $4\frac{1}{2}$

11. $\frac{5}{1}$; $\frac{5}{3}$; $\frac{15}{3}$; $\frac{5}{3}$; $\frac{10}{3}$; $3\frac{1}{3}$

12. $\frac{23}{4}$; $\frac{33}{8}$; $\frac{46}{8}$; $\frac{33}{8}$; $\frac{13}{8}$; $1\frac{5}{8}$

13. $\frac{13}{4}$; $\frac{17}{6}$; $\frac{39}{12}$; $\frac{34}{12}$; $\frac{5}{12}$

14. $1\frac{3}{5}$ 15. $1\frac{3}{7}$

16. $4\frac{3}{10}$ 17. $1\frac{1}{5}$

18. $1\frac{4}{15}$ 19. $1\frac{5}{6}$

Challenge

Magic Square #1

$4\frac{1}{3}$	$9\frac{3}{4}$	$2\frac{1}{6}$
$3\frac{1}{4}$	$5\frac{5}{12}$	$7\frac{7}{12}$
$8\frac{2}{3}$	$1\frac{1}{12}$	$6\frac{1}{2}$

$16\frac{1}{4}$

$8\frac{1}{8}$

Magic Square #2

$2\frac{1}{6}$	$4\frac{7}{8}$	$1\frac{1}{12}$
$1\frac{5}{8}$	$2\frac{17}{24}$	$3\frac{19}{34}$
$4\frac{1}{3}$	$\frac{13}{24}$	$3\frac{1}{4}$

$8\frac{1}{8}$

Magic Square #3

6	$13\frac{1}{2}$	3
$4\frac{1}{2}$	$7\frac{1}{2}$	$10\frac{1}{2}$
12	$1\frac{1}{2}$	9

$22\frac{1}{2}$

$16\frac{7}{8}$

Magic Square #4

$1\frac{1}{2}$	$3\frac{3}{8}$	$\frac{3}{4}$
$1\frac{1}{8}$	$1\frac{7}{8}$	$2\frac{5}{8}$
3	$\frac{3}{8}$	$2\frac{1}{4}$

$5\frac{5}{8}$

Problem Solving

1. $1\frac{3}{4}$ feet 2. $10\frac{13}{20}$ feet

3. $3\frac{3}{4}$ tons 4. $7\frac{1}{12}$ feet

5. A 6. H

7. C 8. J

Reading Strategies

1. fractions 2. integers

3. yes

4. subtract fractions

5. subtract integers

6. $3\frac{2}{4}$, or $3\frac{1}{2}$

Puzzles, Twisters & Teasers

1. R
2. E
3. A
4. C
4. H
6. H
5. I
8. S
9. H
10. A
11. N
12. D
13. S

TO REACH HIS HANDS

LESSON 3-9

Practice A

1. D
2. H
3. D
4. G
5. 2
6. 2
7. 2
8. $\frac{1}{8}$
9. $-\frac{1}{6}$
10. $\frac{1}{2}$
11. –12
12. 20
13. 12
14. $1\frac{1}{8}$
15. 2
16. $1\frac{1}{3}$
17. $1\frac{4}{5}$
18. $3\frac{17}{20}$
19. $7\frac{1}{7}$
20. 3 hours

Practice B

1. $2\frac{1}{2}$
2. $6\frac{3}{4}$
3. $-2\frac{2}{5}$
4. $\frac{3}{7}$
5. $-\frac{1}{2}$
6. $\frac{1}{4}$
7. $-5\frac{1}{7}$
8. $12\frac{1}{2}$
9. $7\frac{7}{8}$
10. $1\frac{2}{5}$
11. $1\frac{7}{10}$
12. $1\frac{14}{15}$
13. $1\frac{3}{5}$
14. $1\frac{1}{2}$
15. $3\frac{1}{2}$
16. $7\frac{7}{8}$
17. $10\frac{2}{3}$
18. $11\frac{3}{5}$
19. $4\frac{1}{2}$ hours
20. $1\frac{1}{5}$ T

Practice C

1. $1\frac{5}{7}$
2. $3\frac{3}{4}$
3. $-1\frac{2}{5}$
4. $-\frac{2}{5}$
5. $\frac{7}{45}$
6. $\frac{2}{3}$
7. $\frac{7}{45}$
8. $\frac{1}{18}$
9. $\frac{1}{8}$
10. $9\frac{1}{2}$
11. $14\frac{17}{20}$
12. $-\frac{11}{12}$
13. $1\frac{5}{11}$
14. $\frac{8}{15}$
15. $16\frac{4}{5}$
16. 2
17. 6
18. 5
19. 2
20. 3
21. 4
22. $41\frac{1}{7}$ years
23. $10\frac{2}{5}$ teaspoons

Review for Mastery

1. $\frac{6}{9} = \frac{2}{3}$

2. $\frac{4 \cdot 5}{5 \cdot 7}; \frac{20}{35} = -\frac{4}{7}$

3. $\frac{10 \cdot 9}{3} = \frac{90}{3} = 30$

4. $\frac{3 \cdot 5}{10 \cdot 2} = \frac{15}{20} = \frac{3}{4}$

5. $\frac{1}{4}$

6. $-\frac{5}{12}$

7. $-\frac{3}{5}$

8. $1\frac{3}{4}$

9. $2\frac{1}{8}$

10. $-1\frac{1}{6}$

11. $-8\frac{2}{3}$

12. 5

13. $3\frac{2}{3}$

Challenge

1. $\frac{3}{40}$ pound
2. $1\frac{199}{200}$ pounds; $13\frac{1}{8}$ pounds
3. $47\frac{39}{40}$ pounds; $60\frac{3}{5}$ pounds
4. 76 pounds; 160 pounds
5. 900 pounds; 2,500 pounds
6. $\frac{133}{400}$ pounds; $1\frac{1}{20}$ pounds
7. $4\frac{1}{5}$ pounds
8. 8,560 pounds

Problem Solving

1. 3 pages
2. 8 friends
3. 400,000 members
4. 8 hours
5. C
6. G
7. D
8. F

Reading Strategies

1. $\frac{3}{5}$
2. 3
3. $\frac{9}{5}$
4. Add the numerators.
5. $\frac{2}{3}$
6. 4
7. $\frac{2}{3} \times 4 = \frac{8}{3}$

Puzzles, Twisters & Teasers

1. I
2. T
3. W
4. O
5. O
6. D
7. E
8. N
9. G
10. O

IT WOODEN GO

LESSON 3-10

Practice A

1. 10
2. 27
3. 24
4. 4
5. 12
6. 16
7. 3
8. 2
9. $\frac{1}{5}$
10. $\frac{1}{2}$
11. $1\frac{1}{2}$
12. $-\frac{1}{4}$
13. $5\frac{1}{3}$
14. $7\frac{1}{2}$
15. $2\frac{6}{7}$
16. $\frac{5}{7}$
17. $\frac{7}{8}$
18. $-\frac{6}{11}$
19. 8 bottles

Practice B

1. 8
2. $\frac{4}{5}$

3. $\frac{5}{9}$

4. $1\frac{1}{3}$

5. $-\frac{1}{2}$

6. $1\frac{1}{6}$

7. $1\frac{1}{24}$

8. $1\frac{11}{16}$

9. $\frac{7}{9}$

10. $-12\frac{1}{2}$

11. $8\frac{1}{8}$

12. $36\frac{2}{3}$

13. $1\frac{2}{7}$

14. $1\frac{11}{34}$

15. $-2\frac{26}{27}$

16. $1\frac{1}{14}$

17. $-1\frac{2}{5}$

18. $6\frac{7}{18}$

19. 140 burgers

20. 7 tiles

Practice C

1. $38\frac{1}{4}$

2. $32\frac{1}{5}$

3. $61\frac{2}{7}$

4. $-\frac{33}{35}$

5. $1\frac{5}{22}$

6. $2\frac{2}{19}$

7. $\frac{9}{55}$

8. $-\frac{1}{3}$

9. $\frac{1}{2}$

10. $12\frac{3}{14}$

11. $17\frac{16}{45}$

12. $-19\frac{23}{55}$

13. $2\frac{7}{27}$

14. $3\frac{6}{23}$

15. $4\frac{4}{55}$

16. $-\frac{16}{21}$

17. $\frac{6}{17}$

18. $16\frac{1}{2}$

19. 5 lawns

20. 7 shelves

Review for Mastery

1. $\frac{2}{1}$; $\frac{8}{5}$; $1\frac{3}{5}$

2. $\frac{6}{5}$; $\frac{30}{40}$; $\frac{3}{4}$

3. $\frac{5}{2} \div \frac{7}{4}$; $\frac{5}{2}$; $\frac{4}{7}$; $\frac{20}{14}$; $\frac{10}{7}$; $1\frac{3}{7}$

4. $\frac{8}{3} \div \frac{6}{5}$; $\frac{8}{3} \bullet \frac{5}{6}$; $\frac{40}{18}$; $\frac{20}{9}$; $2\frac{2}{9}$

5. 2

6. $2\frac{5}{8}$

7. $\frac{5}{6}$

8. $3\frac{9}{10}$

9. $1\frac{1}{3}$

10. $2\frac{1}{4}$

Challenge

1. 18

2. 20

3. 12

4. 40

5. 3

6. 4

7. 16

8. 7

9. $\frac{1}{10}$

10. 20 years

11. mouse; hippopotamus; 38 years

12. 5 years

Problem Solving

1. about 1,026 steps

2. $5\frac{1}{6}$ minutes

3. $1\frac{2}{3}$ miles

4. $\frac{7}{8}$ inch

5. A

6. G

7. B

8. J

Reading Strategies

1. $2\frac{1}{2}$ ft

2. $\frac{1}{2}$ ft

3. 5

4. 5

5. 5

6. 5

7. The answers are the same.

Puzzles, Twisters & Teasers

1. I
2. T
3. M
4. A
5. K
6. E
7. S
8. C
9. E
10. N
11. T
12. S

IT MAKES CENTS

LESSON 3-11

Practice A

1. D
2. G
3. B
4. F
5. D
6. J
7. $p = \frac{5}{12}$
8. $d = \frac{7}{10}$
9. $y = \frac{1}{8}$
10. $m = 1\frac{1}{9}$
11. $x = 1\frac{1}{4}$
12. $r = \frac{9}{25}$
13. $\frac{3}{4}$ mile

Practice B

1. $t = 1$
2. $g = \frac{1}{2}$
3. $k = \frac{7}{10}$
4. $n = \frac{11}{21}$
5. $j = \frac{1}{9}$
6. $f = \frac{1}{3}$
7. $s = 3$
8. $a = 2\frac{1}{2}$
9. $h = 1\frac{1}{9}$
10. $p = 1\frac{7}{24}$
11. $d = 1\frac{3}{10}$
12. $y = 3\frac{15}{28}$
13. $c = 1\frac{3}{4}$
14. $w = 3\frac{1}{15}$
15. $z = 1\frac{26}{35}$
16. $m = 1\frac{1}{15}$
17. $x = 6\frac{14}{15}$
18. $r = 13\frac{1}{3}$
19. $2\frac{6}{25}$ miles
20. $3\frac{5}{6}$ yards

Practice C

1. $t = \frac{17}{19}$
2. $g = \frac{1}{2}$
3. $k = 1\frac{5}{12}$
4. $n = -\frac{7}{30}$
5. $j = -\frac{7}{36}$
6. $f = \frac{4}{55}$
7. $s = 8\frac{3}{4}$
8. $a = 5\frac{10}{21}$
9. $h = 14\frac{1}{7}$
10. $p = 2\frac{37}{56}$
11. $d = -2\frac{11}{20}$
12. $y = 4\frac{38}{45}$
13. $c = 5\frac{17}{20}$
14. $w = 2\frac{42}{55}$
15. $z = 3\frac{1}{18}$
16. $m = 4\frac{4}{35}$
17. $x = -23\frac{1}{10}$
18. $r = 7\frac{7}{9}$
19. 25 mi/h
20. $\frac{7}{8}$ pound

Review for Mastery

1. $\frac{1}{6}; \frac{1}{6}$

   $\frac{9}{12} + \frac{2}{12}$

   $\frac{11}{12}$

2. $\frac{4}{5}; \frac{4}{5}$

   $\frac{12}{15}$

   $\frac{2}{15}$

3. $t = \frac{7}{8}$

4. $k = 1\frac{1}{8}$

5. $a = 1\frac{3}{10}$

6. $\frac{16}{5}$

$\frac{1}{8}$; $\frac{16}{5} \bullet \frac{1}{8}$

$\frac{16}{5} \bullet \frac{1}{8}$

$\frac{16}{40}$; $\frac{2}{5}$

7. $\frac{3}{2}$; $\frac{3}{2}$

$\frac{3}{2}$

$\frac{15}{12}$; $\frac{5}{4}$; $1\frac{1}{4}$

8. $d = 6\frac{2}{3}$

9. $y = \frac{1}{9}$

10. $s = 3\frac{1}{8}$

Challenge

1. $4\frac{1}{2}$

2. $2\frac{1}{4}$ pounds

3. $\frac{3}{4}x = 2\frac{1}{4}$; $x = 3$ cans

4. $1\frac{1}{4}x = 4\frac{1}{2}$; $x = 3\frac{3}{5}$; You need 4 tubs.

5. $\frac{1}{4} + x = 1\frac{1}{8}$; $x = \frac{7}{8}$ cup

6. $4\frac{1}{2} = \frac{1}{4}x$; $x = 18$ green onions

7. $8 - x = 4\frac{1}{2}$; $x = 3\frac{1}{2}$ cloves left

Problem Solving

1. 16 medals
2. 12 gold medals
3. 18 minutes
4. 24 games
5. C
6. H
7. D
8. G

Reading Strategies

1. Get x by itself on one side of the equation.
2. the same
3. Perform the opposite operation.
4. the same
5. addition
6. Solve.
7. the same

Puzzles, Twisters & Teasers

1. $-\frac{121}{126}$
2. $\frac{4}{5}$
3. $\frac{1}{5}$
4. $1\frac{1}{5}$
5. $\frac{2}{7}$
6. $\frac{1}{4}$
7. $\frac{3}{4}$
8. 40
9. $1\frac{1}{2}$
10. $\frac{9}{10}$

TO GET MORE KARATS